THE FEELING
OF WHAT HAPPENS

Born in Portugal, Antonio Damasio is van Allen
Distinguished Professor and head of the Department
of neurology at the University of Iowa College of
Medicine, and Adjunct Professor at the Salk Institute
in La Jolla, California. He is the author of *Descartes'
Error: Emotion, Reason and the Human Brain* and
Looking for Spinoza: Joy, Sorrow and the Feeling Brain.

ALSO BY ANTONIO DAMASIO

Descartes' Error: Emotion, Reason and the Human Brain
Looking for Spinoza: Joy, Sorrow and the Feeling Brain

'This is a landmark book...unquestionably the best book that has yet been written on the subject of consciousness and the brain...The core chapter of the book – "The Neurology of Consciousness" – is brilliantly integrative and original, among the two or three finest pieces of neurological writing I have ever read...The book will challenge and delight the most sophisticated readers, while rarely leaving the less sophisticated lost or overwhelmed'

Journal of Consciousness Studies

'Damasio does not shy away from large, challenging ideas. It is no small achievement to create a language with which to talk about the biological roots of consciousness...*The Feeling of What Happens* will be much talked about'

Houston Chronicle

'Antonio Damasio is a gifted writer...*The Feeling of What Happens* will change your experience of yourself'

New York Times 'Editors' Choice'
Books of the Year

'This book is marvellous – a major work, of meaning and importance. It is an amazing marriage of poetic intuition and precise investigation'

Peter Brook

'Antonio Damasio is not "merely" a successful career neuroscientist and popularizer...he is the major living figure in his field, possessor of the most profound understanding higher human cognition – in short, a genius. *Descartes' Error* and *The Feeling of What Happens* are essential reading. Although they masquerade as "popular science", they are ground-breaking classics of psychology and neuroscience. These are books to buy, keep and ponder upon. Do so, and you will be ahead of the ruck by at least a decade'

Journal of the Royal Society of Medicine

Antonio Damasio

THE FEELING
OF WHAT HAPPENS

Body, Emotion and the Making
of Consciousness

VINTAGE BOOKS
London

FOR HANNA

Published by Vintage 2000

29

Copyright © Antonio Damasio 1999

First published in Great Britain by Heinemann 2000

Vintage
Random House, 20 Vauxhall Bridge Road, London SW1V 2SA

Addresses for companies within The Random House Group Limited can be found at:
www.randomhouse.co.uk/offices.htm

The Random House Group Limited Reg. No. 954009

A CIP catalogue record for this book
is available from the British Library

ISBN 9780099288763

Penguin Random House is committed to a sustainable future for our business, our readers and our planet. This book is made from Forest Stewardship Council® certified paper.

Printed and bound in Great Britain by Clays Ltd, Elcograf S.p.A.

Or the waterfall, or music heard so deeply
That it is not heard at all, but you are the music
While the music lasts. These are only hints and guesses,
Hints followed by guesses; and the rest
Is prayer, observance, discipline, thought and action.
The hint half guessed, the gift half understood, is Incarnation.

T. S. ELIOT
"Dry Salvages" from *Four Quartets*

The question of who I was consumed me.

I became convinced I should not find the image
                              of the person that I
was: Seconds passed. What rose to the surface in me
plunged out of sight again. And yet I felt
the moment of my first investiture
was the moment I began to represent myself —
the moment I began to live — by degrees — second by
second — unrelentingly — Oh mind what you're doing! —

do you want to be *covered* or do you want to be *seen*? —

And the garment — how it becomes you! — starry
with the eyes of
others,
            weeping —

JORIE GRAHAM
"Notes on the Reality of the Self" from *Materialism*

CONTENTS

Maps • Mysteries and Gaps of Knowledge in the Making of Images •
New Terms • Some Pointers on the Anatomy of the Nervous System •
The Brain Systems behind the Mind

The Feeling
of What Happens

PART I

Introduction

Chapter One

Stepping into the Light

STEPPING INTO THE LIGHT

I have always been intrigued by the specific moment when, as we sit waiting in the audience, the door to the stage opens and a performer steps into the light; or, to take the other perspective, the moment when a performer who waits in semidarkness sees the same door open, revealing the lights, the stage, and the audience.

I realized some years ago that the moving quality of this moment, whichever point of view one takes, comes from its embodiment of an instance of birth, of passage through a threshold that separates a protected but limiting shelter from the possibility and risk of a world beyond and ahead. As I prepare to introduce this book, however, and as I reflect on what I have written, I sense that stepping into the light is also a powerful metaphor for consciousness, for the birth of the knowing mind, for the simple and yet momentous coming of the sense of self into the world of the mental. How we step into the light of consciousness is precisely the topic of this book. I write about the sense of self and about the transition from innocence and ignorance to knowingness and selfness. My specific goal is to consider the biological circumstances that permit this critical transition.

No aspect of the human mind is easy to investigate, and for those who wish to understand the biological underpinnings of the mind, consciousness is generally regarded as the towering problem, in spite of the fact that the definition of the problem may vary considerably from investigator to investigator. If elucidating mind is the last frontier of the life sciences, consciousness often seems like the last mystery in the elucidation of the mind. Some regard it as insoluble.

Yet, it is difficult to think of a more seductive challenge for reflection and investigation. The matter of mind, in general, and of consciousness, in particular, allows humans to exercise, to the vanishing point, the desire for understanding and the appetite for wonderment at their own nature that Aristotle recognized as so distinctively human. What could be more difficult to know than to know how we know? What could be more dizzying than to realize that it is our having consciousness which makes possible and even inevitable our questions about consciousness?

Although I do not see consciousness as the pinnacle of biological evolution, I see it as a turning point in the long history of life. Even when we resort to the simple and standard dictionary definition of consciousness—as an organism's awareness of its own self and surroundings—it is easy to envision how consciousness is likely to have opened the way in human evolution to a new order of creations not possible without it: conscience, religion, social and political organizations, the arts, the sciences, and technology. Perhaps even more compellingly, consciousness is the critical biological function that allows us to know sorrow or know joy, to know suffering or know pleasure, to sense embarrassment or pride, to grieve for lost love or lost life. Whether individually experienced or observed, pathos is a by-product of consciousness and so is desire. None of those personal states would ever be known to each of us without consciousness. Do not blame Eve for knowing; blame consciousness, and thank it, too.

I write this in downtown Stockholm as I look out of a window and watch a frail old man make his way toward a ferry that is about to depart. Time is short, but his gait is slow; his steps break at the ankle from arthritic pain; his hair is white; his coat is worn. It is raining persistently and the wind makes him bend slightly like a lone tree in an open field. He finally reaches the ship. He climbs with difficulty the tall step needed to get on the gangplank and starts on his way down to the deck, afraid of gaining too much momentum on the incline, head moving briskly, left and right, checking his surroundings and seeking reassurance, his whole body seemingly saying, Is this it? Am I in the right place? Where to next? And then the two men on deck help him steady his last step, ease him into the cabin with warm gestures, and he seems to be safely where he should be. My worry is over. The ship departs.

Now let your mind wander and consider that, without consciousness, the old man's discomfort, perhaps humiliation, would simply not have been known to him. Without consciousness, the two men on deck would not have responded with empathy. Without consciousness, I would not have been concerned and would never have thought that one day I might be him, walking with the same pained hesitation and feeling the same discomfort. Consciousness amplifies the impact of these feelings in the minds of the characters in this scene.

Consciousness is, in effect, the key to a life examined, for better and for worse, our beginner's permit into knowing all about the hunger, the thirst, the sex, the tears, the laughter, the kicks, the punches, the flow of images we call thought, the feelings, the words, the stories, the beliefs, the music and the poetry, the happiness and the ecstasy. At its simplest and most basic level, consciousness lets us recognize an irresistible urge to stay alive and develop a concern for the self. At its most complex and elaborate level, consciousness helps us develop a concern for other selves and improve the art of life.

Absent without Leave

Thirty-two years ago, a man sat across from me in a strange, entirely circular, gray-painted examining room. The afternoon sun was shining on us through a skylight as we talked quietly. Suddenly the man stopped, in midsentence, and his face lost animation; his mouth froze, still open, and his eyes became vacuously fixed on some point on the wall behind me. For a few seconds he remained motionless. I spoke his name but there was no reply. Then he began to move a little, he smacked his lips, his eyes shifted to the table between us, he seemed to see a cup of coffee and a small metal vase of flowers; he must have, because he picked up the cup and drank from it. I spoke to him again and again he did not reply. He touched the vase. I asked him what was going on, and he did not reply, his face had no expression. He did not look at me. Now, he rose to his feet and I was nervous; I did not know what to expect. I called his name and he did not reply. When would this end? Now he turned around and walked slowly to the door. I got up and called him again. He stopped, he looked at me, and some expression returned to his face—he looked perplexed. I called him again, and he said, "What?"

For a brief period, which seemed like ages, this man suffered from an impairment of consciousness. Neurologically speaking, he had an absence seizure followed by an absence automatism, two among the many manifestations of epilepsy, a condition caused by brain dysfunction. This was not my first exposure to impaired consciousness but it was the most intriguing yet. From a first-person perspective, I knew what it was like to dissolve into unsolicited unknowingness and to return to consciousness—I had lost consciousness once, as a kid, in an accident, and I had general anesthesia once, as an adolescent. I also had seen patients in coma and observed, from a third-person perspective, what a state of unconsciousness looked like. In all of these instances, however, as well as in falling asleep or waking up, the loss of consciousness was radical, something like a complete power outage. But what I had just seen that afternoon in the gray circular room was far more startling. The man had not collapsed on the floor, comatose, and had not gone to sleep, either. He was both there and not there, certainly awake, attentive in part, behaving for sure, bodily present but personally unaccounted for, absent without leave.

This episode stayed with me and it was a good day when I felt I could interpret its meaning. I did not think then, but I think now, that I had witnessed the razor-sharp transition between a fully conscious mind and a mind deprived of the sense of self. During the period of impaired consciousness, the man's wakefulness, his basic ability to attend to objects, and his capacity to navigate in space had been preserved. The essence of his mental process was probably retained, as far as the objects in his surroundings were concerned, but his sense of self and knowing had been suspended. The shaping of my notion of consciousness probably began that day, without my noticing it, and the idea that a sense of self was an indispensable part of the conscious mind only gained strength as I saw comparable cases.

I maintained an interest in the issue of consciousness through the years, at once attracted to the scientific challenge posed by consciousness and repulsed by the human consequences of its impairment in neurological disease, but I remained distant. The drama of the situations in which brain damage causes coma or persistent vegetative state, the conditions in which consciousness is most radically impaired, is something I would have preferred not to observe, if given a choice. Few things are as sad to watch as the sudden and forced disappearance of the conscious mind in someone who remains alive, and few things are as painful to explain to a family. How does one look a person in the eye and make clear that the quiet state of a lifetime's companion may appear like sleep but is not sleep; that there is nothing benign or restorative about this way of resting; that the once-sentient being may never return to sentience? But even if my life as a neurologist had not made me wary of consciousness, my life as a neuroscientist might have ensured I did not touch the problem. Studying consciousness was simply not the thing to do before you made tenure, and even after you did it was looked upon with suspicion. Only in recent years has consciousness become a somewhat safer topic of scientific inquiry.[1]

Still, the reason why I eventually turned to consciousness had little to do with the sociology of consciousness studies. I certainly had not planned on investigating consciousness until an impasse forced me to do so. The impasse had to do with my work on the emotions, and that means I can blame the consequences on the passions of the soul.[2]

So here is the situation. I could understand reasonably well how different emotions were induced in the brain and played out in the theater of the body. I could also envision how both the induction of emotions and the consequent bodily changes that largely constitute an emotional state were signaled in several brain structures appropriate to map such changes, thus constituting the substrate for feeling an emotion. But, for the life of me, I could not understand how that brain substrate of feeling could become *known* to the organism having the emotion. I could not devise a satisfactory explanation for how what we conscious creatures call feeling becomes known to the feeling organism. By which additional mechanism do each of us know that a feeling is occurring within the bounds of our own organism? What else happens in the organism and, especially, what else happens in the brain, when we know that we feel an emotion or feel pain or, for that matter, when we know anything at all? I had come up against the obstacle of consciousness. Specifically, I had come up against the obstacle of self, for something like a sense of self was needed to make the signals that constitute the feeling of emotion known to the organism having the emotion.

I could see that overcoming the obstacle of self, which meant, from my standpoint, understanding its neural underpinnings, might help us understand the very different biological impact of three distinct although closely related phenomena: *an emotion, the feeling of that emotion,* and *knowing that we have a feeling of that emotion.* No less important, overcoming the obstacle of self might also help elucidate the neural underpinnings of consciousness in general.

THE PROBLEM OF CONSCIOUSNESS

What is the problem of consciousness, then, from the perspective of neurobiology? Much as I see the matter of self as a critical issue in the elucidation of consciousness, it is important to make clear that the problem of consciousness is not confined to the matter of self. In the simplest of summaries, I regard the problem of consciousness as a combination of two intimately related problems. The first is the problem of understanding how the brain inside the human organism engenders the mental patterns we call, for lack of a better term, the images of an object. By *object* I mean entities as diverse as a person, a place, a melody, a toothache, a state of bliss; by *image* I mean a mental pattern in any of the sensory modalities, e.g., a sound image, a tactile image, the image of a state of well-being. Such images convey aspects of the physical characteristics of the object and they may also convey the reaction of like or dislike one may have for an object, the plans one may formulate for it, or the web of relationships of that object among other objects. Quite candidly, this first problem of consciousness is the problem of how we get a "movie-in-the-brain," provided we realize that in this rough metaphor the movie has as many sensory tracks as our nervous system has sensory portals—sight, sound, taste, and olfaction, touch, inner senses, and so on. (See the glossary section of the appendix for a comment on the use of terms such as *image, representation,* and *map.*)

From the perspective of neurobiology, solving this first problem consists in discovering how the brain makes neural patterns in its nerve-cell circuits and manages to turn those neural patterns into the explicit mental patterns which constitute the highest level of biological phenomenon, which I like to call images. Solving this problem encompasses, of necessity, addressing the philosophical issue of qualia. Qualia are the simple sensory qualities to be found in the blueness of the sky or the tone of sound produced by a cello, and the fundamental components of the images in the movie metaphor are thus made of qualia. I believe these qualities will be eventually explained neurobiologically although at the moment the neurobiological account is incomplete and there is an explanatory gap.[3]

Now, for the second problem of consciousness. This is the problem of how, in parallel with engendering mental patterns for an object, the brain also engenders a sense of self in the act of knowing. To help me clarify what I mean by *self* and *knowing*, I urge you to check their presence in your own mind right now.

You are looking at this page, reading the text and constructing the meaning of my words as you go along. But concern with text and meaning hardly describes all that goes on in your mind. In parallel with representing the printed words and displaying the conceptual knowledge required to understand what I wrote, your mind also displays something else, something sufficient to indicate, moment by moment, that *you* rather than anyone else are doing the reading and the understanding of the text. The sensory images of what you perceive externally, and the related images you recall, occupy most of the scope of your mind, but not all of it. Besides those images there is also this other presence that signifies you, as observer of the things imaged, owner of the things imaged, potential actor on the things imaged. There is a presence of you in a particular relationship with some object. If there were no such presence, how would your thoughts belong to you? Who could tell that they did? The presence is quiet and subtle, and sometimes it is little more than a "hint half guessed," a "gift half understood," to borrow words from T. S. Eliot. Later I shall propose that the simplest form of such a presence is also an image, actually the kind of image that constitutes a feeling. In that perspective, the presence of you is the feeling of what happens when your being is modified by the acts of apprehending something. The presence never quits, from the moment of awakening to the moment sleep begins. The presence must be there or there is no you.

The solution for this second problem requires the understanding of how, as I write, I have a sense of me, and how, as you now read, you have a sense of you; of how we sense that the proprietary knowledge you and I behold in our minds, this very moment, is shaped in a particular perspective, that of the individual inside of whom it is formed, rather than in some canonical, one-type-fits-all perspective. The solution also requires the understanding of how the images of an object and of the complex matrix of relations, reactions, and plans related to it are sensed as the unmistakable mental property of an automatic owner who, for all intents and purposes, is an observer, a perceiver, a knower, a thinker, and a potential actor. This second problem is all the more intriguing since we can be certain that the solution traditionally proposed for it—a homunculus creature who is in charge of knowing—is patently incorrect. There is no homunculus, either metaphysical or in the brain, sitting in the Cartesian theater as an audience of one and waiting for objects to step into the light.[4] In other words, solving the second problem of consciousness consists in discovering the biological underpinnings for the curious ability we humans have of constructing, not just the mental patterns of an object—the images of persons, places, melodies, and of their relationships, in short, the temporally and spatially integrated mental images of something-to-be-known—but also the mental patterns which convey, automatically and naturally, the sense of a self in the act of knowing. Consciousness, as we commonly think of it, from its basic levels to its most complex, is the unified mental pattern that brings together the object and the self.

In the very least, then, the neurobiology of consciousness faces two problems: the problem of how the movie-in-the-brain is generated, and the problem of how the brain also generates the sense that there is an owner and observer for that movie. The two problems are so intimately related that the latter is nested within the former. In effect, the second problem is that of generating the *appearance* of an owner and observer for the movie *within the movie*; and the physiological mechanisms behind the second problem have an influence on the mechanisms behind the first. In spite of the intimacy of the problems, however, separating them is a way of breaking the problem of consciousness into parts and, in so doing, making the overall investigation of consciousness manageable.[5]

This book is about an attempt to deal with the obstacle of consciousness focusing squarely on the problem of self but neither neglecting nor minimizing the "other" problem of consciousness. The attempt was prompted by the impasse on emotions described earlier, but it has gone beyond addressing that particular issue. The book is about my idea of what consciousness is, in mental terms, and about how consciousness can be constructed in the human brain. I do not claim to have solved the problem of consciousness, and at the current stage in the history of cognitive science and neuroscience, notwithstanding several new and substantial contributions, I regard the thought of solving *the* consciousness problem with some skepticism. I simply hope that the ideas presented here help with the eventual elucidation of the problem of self from a biological perspective.[6]

The background for the text is an ongoing research program that relies on varied lines of investigative activity—reflecting on facts gleaned from many years of observation of neurological patients with disorders of mind and behavior and on findings from experimental neuropsychological studies of such disorders; theorizing about the processes of consciousness as they occur in the normal human condition, using evidence from general biology, neuroanatomy, and neurophysiology; and designing testable hypotheses regarding the neuroanatomical underpinnings of consciousness informed by reflection and theory.

APPROACHING CONSCIOUSNESS

Before we go any farther it is necessary to say a few words about how to approach the problem we have defined. It would be wonderful, of course, if the contents of our minds were even more richly superposed than they already are, so that I could write this book in parallel tracks and you could read, simultaneously, about theoretical assumptions, scientific methods, and foundational facts. But we operate in a world of classical physics and I must resort to devices of the Elizabethan age: asides and digressions. I promise to be brief and stick to the essentials.

Mind, Behavior, and Brain

Consciousness is an entirely private, first-person phenomenon which occurs as part of the private, first-person process we call mind.[7] Consciousness and mind, however, are closely tied to external behaviors that can be observed by third persons. We all share these phenomena—mind, consciousness within mind, and behaviors—and we know quite well how they are intercorrelated, first because of our own self-analysis, second because of our natural propensity to analyze others. Both wisdom and the science of the human mind and behavior are based on this incontrovertible correlation between the private and the public—first-person mind, on the one hand, and third-person behavior, on the other. Fortunately, for those of us who also wish to understand the mechanisms behind mind and behavior, it so happens that mind and behavior are also closely correlated with the functions of living organisms, specifically with the functions of the brain within those organisms.[8] The power of this triangulation of mind, behavior, and brain has been apparent for over a century and a half—ever since the neurologists Paul Broca and Carl Wernicke discovered a connection between language and certain regions of the left cerebral hemisphere. The triangulation has allowed a most felicitous development: the traditional worlds of philosophy and psychology have gradually joined forces with the world of biology and created an odd but productive alliance. For example, by means of the loose federation of scientific approaches currently known as cognitive neuroscience, the alliance has permitted new advances in the understanding of vision, memory, and language. There is good reason to expect that the alliance will assist with the understanding of consciousness as well.

Over the past two decades, work in cognitive neuroscience has become especially rewarding because the development of new techniques to observe the brain in terms of its structure and function now permits us to link a certain behavior we observe, clinically or in an experiment, not only to the presumed mental counterpart of that behavior, but also to specific indices of brain structure or brain activity.

Let me offer some examples. Areas of circumscribed brain damage caused by neurological disease, which are known as lesions, have long been a mainstay of research on the neural basis of the mind. Such lesions used to be revealed only at the time of autopsy, often many years after the study of the patient had been concluded. This time lag slowed the process of analysis and generated some uncertainty in the correlation between anatomy and behavior. Recent technical developments, however, permit us to analyze the lesions in a 3-D reconstruction of the living patient's brain at the same time behavioral or cognitive observations are being carried out. The reconstruction is displayed on a computer screen and is based on an elaborate manipulation of raw data obtained from a magnetic resonance scan. It depicts neural structures with great fidelity and allows careful dissection in virtual space rather than on a laboratory bench. The significance of this development is that a lesion analyzed in this detailed and timely manner serves as a probe to test hypotheses about how a brain system performs a certain mental function or behavior. For instance, we may postulate that a system made up of four interconnected brain regions, A, B, C, D, operates in a particular fashion. Then we may predict the kind of changes that must occur when, say, region C is destroyed. To test the validity of the prediction we study how patients with a lesion in area C behave while performing a given task. Incidentally, the same approach is used in another recently evolved area of neuroscience, molecular neurobiology. A specific gene is inactivated experimentally, in a mouse, for instance, thus causing a "lesion" (in scientific jargon this is called a "knock-out"). The investigators can then determine whether the consequences of the "knock-out" are as predicted.[9]

Another example of a new type of brain index is an area of increased or decreased brain activity revealed by a positron emission scan (PET) or a functional magnetic resonance imaging scan (fMRI). Such scans can be used not only in neurological patients but also in humans without brain diseases. Again, a specific prediction concerning the activity of a certain region during the performance of a particular mental task is used to assess the validity of the hypothesis.

Yet another index is a change in electrical conductance response measured in the skin; or a change in electrical potentials and related magnetic fields measured from the scalp; or a change in electrical potentials measured directly on the brain surface during surgery for epilepsy. Remarkably, the possibility of making intricate linkages among private mind, public behavior, and brain function does not stop with the application of these new techniques. The cross linkages can be extended by a connection to new domains of knowledge about the anatomy and function of the nervous system, gathered by experimental neuroanatomists, neurophysiologists, neuropharmacologists, and neurobiologists who study molecular events within individual nerve cells and can, in turn, relate those events to the composition and action of specific genes. The facts gathered recently on the basis of all these developments allow us to establish progressively more detailed theories regarding the relation between certain aspects of mind and behavior and the brain. The organism's private mind, the organism's public behavior, and its hidden brain can thus be joined in the adventure of theory, and out of the adventure come hypotheses that can be tested experimentally, judged on their merits, and subsequently endorsed, rejected, or modified. (See the appendix for fundamentals of brain anatomy and organization.)

Reflecting on the Neurological and Neuropsychological Evidence

The results of neurological observations and of neuropsychological experiments reveal many facts that were the starting point for the ideas presented here. The first fact is that some aspects of the processes of consciousness can be related to the operation of specific brain regions and systems, thus opening the door to discovering the neural architecture which supports consciousness. The regions and systems in question cluster in a limited set of brain territories and no less so than with functions such as memory or language there will be an anatomy of consciousness. One of the purposes of this text is to present testable anatomical hypotheses for some aspects of the consciousness process.

The second fact is that consciousness and wakefulness, as well as consciousness and low-level attention, can be separated. This fact was based on the evidence that patients can be awake and attentive without having normal consciousness, as exemplified by the man in the circular room. In chapters 3 and 4, I discuss such patients and consider the theoretical significance of their conditions.

The third and perhaps most revealing fact is that consciousness and emotion are *not* separable. As discussed in chapters 2, 3, and 4, it is usually the case that when consciousness is impaired so is emotion. In effect, the connection between emotion and consciousness, on the one hand, and between both of these and the body, on the other, form a main theme of this book.

The fourth fact is that consciousness is not a monolith, at least not in humans: it can be separated into simple and complex kinds, and the neurological evidence makes the separation transparent. The simplest kind, which I call *core consciousness,* provides the organism with a sense of self about one moment—now—and about one place—here. The scope of core consciousness is the here and now. Core consciousness does not illuminate the future, and the only past it vaguely lets us glimpse is that which occurred in the instant just before. There is no elsewhere, there is no before, there is no after. On the other hand, the complex kind of consciousness, which I call *extended consciousness* and of which there are many levels and grades, provides the organism with an elaborate sense of self—an identity and a person, you or me, no less—and places that person at a point in individual historical time, richly aware of the lived past and of the anticipated future, and keenly cognizant of the world beside it.

In short, core consciousness is a simple, biological phenomenon; it has one single level of organization; it is stable across the lifetime of the organism; it is not exclusively human; and it is not dependent on conventional memory, working memory, reasoning, or language. On the other hand, extended consciousness is a complex biological phenomenon; it has several levels of organization; and it evolves across the lifetime of the organism. Although I believe extended consciousness is also present in some nonhumans, at simple levels, it only attains its highest reaches in humans. It depends on conventional memory and working memory. When it attains its human peak, it is also enhanced by language.

The supersense of core consciousness is the first step into the light of knowing and it does not illuminate a whole being. On the other hand, the supersense of extended consciousness eventually brings a full construction of being into the light. In extended consciousness, both the past and the anticipated future are sensed along with the here and now in a sweeping vista as far-ranging as that of an epic novel.

If it is true that core consciousness is the rite of passage into knowing, it is equally true that the levels of knowing which permit human creativity are those which only extended consciousness allows. When we think of the glory that is consciousness, and when we consider consciousness as distinctively human, we are thinking of extended consciousness at its zenith. And yet, as we shall see, extended consciousness is not an independent variety of consciousness: on the contrary, it is built on the foundation of core consciousness. The fine scalpel of neurological disease reveals that impairments of extended consciousness allow core consciousness to remain unscathed. By contrast, impairments that begin at the level of core consciousness demolish the entire edifice of consciousness: extended consciousness collapses as well. The glory that is consciousness requires the orderly enhancement of both kinds of consciousness. But if we are to elucidate the glorious combination, we are well advised to begin by understanding the simpler, foundational kind: core consciousness.[10]

Incidentally, the two kinds of consciousness correspond to two kinds of self. The sense of self which emerges in core consciousness is the *core self*, a transient entity, ceaselessly re-created for each and every object with which the brain interacts. Our traditional notion of self, however, is linked to the idea of identity and corresponds to a nontransient collection of unique facts and ways of being which characterize a person. My term for that entity is the *autobiographical self*. The autobiographical self depends on systematized memories of situations in which core consciousness was involved in the knowing of the most invariant characteristics of an organism's life—who you were born to, where, when, your likes and dislikes, the way you usually react to a problem or a conflict, your name, and so on. I use the term *autobiographical memory* to denote the organized record of the main aspects of an organism's biography. The two kinds of self are related, and in chapter 6, I explain how the autobiographical self arises from the core self.

A fifth fact: not infrequently, consciousness is simply explained in terms of other cognitive functions, such as language, memory, reason, attention, and working memory. While such functions are indeed necessary for the top tiers of extended consciousness to operate normally, the study of neurological patients suggests that they are not required for core consciousness. Accordingly, a theory of consciousness should *not* be just a theory of how memory, reason, and language help construct, from the top down, an interpretation of what goes on in the brain and mind. To be sure, memory, intelligent inferences, and language are critical to the generation of what I call the autobiographical self and the process of extended consciousness. Some interpretation of the events that take place in an organism can surely arise after the process of autobiographical self and extended consciousness are in place. But I do not believe consciousness began that way, at that high a level in the hierarchy of cognitive processes and that late in the history of life and of each of us. I propose that the earliest forms of consciousness precede inferences and interpretations—they are part of the biological transition that eventually enables inferences and interpretations. Accordingly, a theory of consciousness should account for the simpler, foundational kind of the phenomenon which occurs close to the nonconscious representation of the organism for whose sake the entire show is put together and which can support the later developments of identity and person.

Moreover, a theory of consciousness should *not* be just a theory of how the brain attends to the image of an object. As I see it, natural low-level attention precedes consciousness, while focused attention follows the unfolding of consciousness. Attention is as necessary to consciousness as having images. But attention is not sufficient for consciousness and is not the same as consciousness.

Finally, a theory of consciousness should *not* be just a theory of how the brain creates integrated and unified mental scenes, although the production of integrated and unified mental scenes is an important aspect of consciousness, especially at its highest levels. Those scenes do not exist in a vacuum. I believe they are integrated and unified *because* of the singularity of the organism and *for* the benefit of that single organism. The mechanisms that prompt the integration and unification of the scene require an explanation.

By focusing the explanatory efforts on how the sense of self in the act of knowing an object appears in the mind, I am open to the criticism that I am *just* addressing the problem of so-called self-consciousness and neglecting the remainder of the problem, namely the qualia problem. I would answer the criticism as follows. If "self-consciousness" is taken to mean "consciousness with a sense of self," then all human consciousness is necessarily covered by the term—there is just no other kind of consciousness as far as I can see. I would add that the biological state we describe as sense of self and the biological machinery responsible for engendering it may well have a hand in optimizing the processing of the objects to be known—having a sense of self is not only required for knowing, in the proper sense, but may influence the processing of whatever gets to be known. In other words, the biological processes that pose the second problem of consciousness probably plays a role in the biological processes that pose the first. When I address the self problem, I address the qualia issue with respect to the representation of the organism having consciousness.[11]

A SEARCH FOR SELF

How do we ever know that we are seeing a given object? How do we become conscious in the full sense of the word? How is the sense of self in the act of knowing implanted in the mind? The way into a possible answer for the questions on self came only after I began seeing the problem of consciousness in terms of two key players, the *organism* and the *object,* and in terms of the *relationships* those players hold in the course of their natural interactions. The organism in question is that within which consciousness occurs; the object in question is any object that gets to be known in the consciousness process; and the relationships between organism and object are the contents of the knowledge we call consciousness. Seen in this perspective, consciousness consists of constructing knowledge about two facts: that the organism is involved in relating to some object, and that the object in the relation causes a change in the organism.

The new perspective also makes the biological realization of consciousness a treatable problem. The process of knowledge construction requires a brain, and it requires the signaling properties with which brains can assemble neural patterns and form images. The neural patterns and images necessary for consciousness to occur are those which constitute proxies for the organism, for the object, and for the relationship between the two. Placed in this framework, understanding the biology of consciousness becomes a matter of discovering how the brain can map *both* the two players *and* the relationships they hold.

The general problem of representing the object is not especially enigmatic. Extensive studies of perception, learning and memory, and language have given us a workable idea of how the brain processes an object, in sensory and motor terms, and an idea of how knowledge about an object can be stored in memory, categorized in conceptual or linguistic terms, and retrieved in recall or recognition modes. The neurophysiologic details of these processes have not been worked out, but the contours of these problems are understandable. From my perspective, neuroscience has been dedicating most of its efforts to understanding the neural basis of what I see as the "object proxy." In the relationship play of consciousness, the object is exhibited in the form of neural patterns in the sensory cortices appropriate to map its characteristics. For example, in the case of the visual aspects of an object, the neural patterns are constructed in a variety of regions of the visual cortices, not just one or two, but many, working in concerted fashion to map the varied aspects of the object in visual terms.[12] On the side of the organism, however, matters are quite different. To indicate how different matters are, let me suggest an exercise.

Look up from the page, at whatever is directly in front of you, observe intently, and then return to the page. As you did so, the many stations of your visual system, from the retinas to several regions of the brain's cerebral cortex, shifted rapidly from mapping the book's page, to mapping the room in front of you, to mapping the page again. Now turn around 180 degrees and look at what is behind you. Again, mapping of the page vanished swiftly so that the visual system could map the new scene you were contemplating. The moral of the story: in quick succession, precisely the *same* brain regions constructed several entirely *different* maps by virtue of the different motor settings the organism assumed and of the different sensory inputs the organism gathered. The image constructed in the brain's multiplex screens changed remarkably.

Now consider this: while your visual system changed dutifully at the mercy of the objects it mapped, a number of regions in your brain whose job it is to regulate the life process and which contain preset maps that represent varied aspects of your body did not change at all in terms of the *kind* of object they represented. The body remained the "object" all along and will remain so until death ensues. But not only was the *kind* of object precisely the same; the degree of change occurring in the object—the body—was quite small. Why was that so? Because only a narrow range of body states is compatible with life, and the organism is genetically designed to maintain that narrow range and equipped to seek it, through thick and through thin.

What we have in this situation, then, is an intriguing asymmetry that may be phrased in the following terms: some parts of the brain are free to roam over the world and in so doing are free to map whatever object the organism's design permits them to map. On the other hand, some other parts of the brain, those that represent the organism's own state, are not free to roam at all. They are stuck. They can map nothing but the body and do so within largely preset maps. They are the body's captive audience, and they are at the mercy of the body's dynamic sameness.

There are several reasons behind this asymmetry. First, the composition and general functions of the living body remain the same, in terms of their quality, across a lifetime. Second, the body changes that continuously do occur are small, in terms of their quantity. They have a narrow dynamic range because the body must operate with a limited range of parameters if it is to survive; the body's internal state must be relatively stable by comparison to the environment surrounding it. Third, that stable state is governed from the brain by means of an elaborate neural machinery designed to detect minimal variations in the parameters of the body's internal chemical profile and to command actions aimed at correcting the detected variations, directly or indirectly. (I will address the neuroanatomy of this system in chapter 5. The system is made of not one but many units, the most important of which are located in the brain stem, hypothalamus, and basal forebrain sections of the brain.) In short, the organism in the relationship play of consciousness is the entire unit of our living being, our body as it were; and yet, as it turns out, the part of the organism called the brain holds within it a sort of model of the whole thing. This is a strange, overlooked, and noteworthy fact, and is perhaps the single most important clue as to the possible underpinning of consciousness.

I have come to conclude that the organism, as represented inside its own brain, is a likely biological forerunner for what eventually becomes the elusive sense of self. The deep roots for the self, including the elaborate self which encompasses identity and personhood, are to be found in the ensemble of brain devices which continuously and *nonconsciously* maintain the body state within the narrow range and relative stability required for survival. These devices continually represent, *nonconsciously,* the state of the living body, along its many dimensions. I call the state of activity within the ensemble of such devices the *proto-self,* the nonconscious forerunner for the levels of self which appear in our minds as the conscious protagonists of consciousness: core self and autobiographical self.

Should some readers get worried, at this point, that I am falling into the abyss of the homunculus trap, let me say immediately and vehemently that that is not the case. The "model of the body-in-the-brain" to which I am referring is nothing at all like the rigid homunculus creature of old-fashioned neurology textbooks. Nothing in it looks like a little person inside a big person; the model "perceives" nothing and "knows" nothing; it does not talk and it does not make consciousness. The model is, instead, a collection of brain devices whose main job is the automated management of the organism's life. As we shall discuss, the management of life is achieved by a variety of innately set regulatory actions—secretion of chemical substances such as hormones as well as actual movements in viscera and in limbs. The deployment of these actions depends on the information provided by nearby neural maps which signal, moment by moment, the state of the entire organism. Most importantly, neither the life-regulating devices nor their body maps are the generators of consciousness, although their presence is indispensable for the mechanisms that do achieve core consciousness.

This is the key issue, as argued in chapter 5: in the relationship play of consciousness, the organism is represented in the brain, abundantly and multifariously, and that representation is tied to the maintenance of the life process. If this idea is correct, life and consciousness, specifically the self aspect of consciousness, are indelibly interwoven.

WHY WE NEED CONSCIOUSNESS

If you find the connection between life and consciousness surprising, consider the following. Survival depends on finding and incorporating sources of energy and on preventing all sorts of situations which threaten the integrity of living tissues. It is certainly true that without actions organisms such as ours would not survive since the sources of energy required for renewing the organism's structure and maintaining life would not be found and harnessed to the service of the organism, never mind staving off environmental dangers. But on their own, without the guidance of images, actions would not take us far. Good actions need the company of good images. Images allow us to choose among repertoires of previously available patterns of action and optimize the delivery of the chosen action—we can, more or less deliberately, more or less automatically, review mentally the images which represent different options of action, different scenarios, different outcomes of action. We can pick and choose the most appropriate and reject the bad ones. Images also allow us to invent new actions to be applied to novel situations and to construct plans for future actions—the ability to transform and combine images of actions and scenarios is the wellspring of creativity.

If actions are at the root of survival and if their power is tied to the availability of guiding images, it follows that a device capable of maximizing the effective manipulation of images in the service of the interests of a particular organism would have given enormous advantages to the organisms that possessed the device and would probably have prevailed in evolution. Consciousness is precisely such a device.

The pathbreaking novelty provided by consciousness was the possibility of connecting the inner sanctum of life regulation with the processing of images. Put in other words, it was the possibility of bringing the system of life regulation—which is housed in the depths of the brain in regions such as the brain stem and hypothalamus—to bear on the processing of the images which represent the things and events which exist inside and outside the organism. Why was this really an advantage? Because survival in a complex environment, that is, efficient management of life regulation, depends on taking the right action, and that, in turn, can be greatly improved by purposeful preview and manipulation of images in mind and optimal planning. Consciousness allowed the connection of the two disparate aspects of the process—inner life regulation and image making.

Consciousness generates the knowledge that images exist within the individual who forms them, it places images in the organism's perspective by referring those images to an integrated representation of the organism, and, in so doing, allows the manipulation of the images to the organism's advantage. Consciousness, when it appears in evolution, announces the dawn of individual forethought.

Consciousness opens the possibility of constructing in the mind some counterpart to the regulatory specifications hidden in the brain core, a new way for the life urge to press its claims and for the organism to act on them. Consciousness is the rite of passage which allows an organism armed with the ability to regulate its metabolism, with innate reflexes, and with the form of learning known as conditioning, to become a minded organism, the kind of organism in which responses are shaped by a mental *concern* over the organism's own life. Spinoza said that the effort to preserve oneself is the first and unique foundation of virtue.[13] Consciousness enables that effort.

THE BEGINNING OF CONSCIOUSNESS

Once I could envision how the brain might put together the patterns that stand for an object and those that stand for the organism, I began considering the mechanisms that the brain may use to represent the relationship between object and organism. I was looking specifically for how the brain might represent the fact that when an organism is engaged in the processing of an object, the object *causes* the organism to react and, in so doing, change its state. A possible solution is presented in chapters 6, 7, and 8. I propose that we become conscious when the organism's representation devices exhibit a specific kind of wordless knowledge—the knowledge that the organism's own state has been changed by an object—and when such knowledge occurs along with the salient representation of an object. The sense of self in the act of knowing an object is an infusion of *new* knowledge, continuously created within the brain as long as "objects," actually present or recalled, interact with the organism and cause it to change.

The sense of self is the first answer to a question the organism never posed: To whom do the ongoing mental patterns now unfolding belong? The answer is that they belong to the organism, as represented by the proto-self. Later I indicate how the brain assembles the wordless knowledge necessary to produce this unrequested answer. At this point, however, I can say that the simplest form in which the wordless knowledge emerges mentally is the feeling of knowing—the feeling of what happens when an organism is engaged with the processing of an object—and that only thereafter can inferences and interpretations begin to occur regarding the feeling of knowing.

In a curious way, consciousness begins as the feeling of what happens when we see or hear or touch. Phrased in slightly more precise words, it is a feeling that accompanies the making of any kind of image—visual, auditory, tactile, visceral—within our living organisms. Placed in the appropriate context, the feeling marks those images as ours and allows us to say, in the proper sense of the terms, that we see or hear or touch. Organisms unequipped to generate core consciousness are condemned to making images of sight or sound or touch, there and then, but cannot come to know that they did. From its most humble beginnings, consciousness is knowledge, knowledge consciousness, no less interconnected than truth and beauty were for Keats.

COPING WITH MYSTERY

There has been a lack of agreement among those studying the problem of consciousness, not only about what consciousness is but also about the prospects of understanding its biological underpinnings. There has also been some degree of puzzlement, and even worry, among those who are not scholars of consciousness but simple day-to-day users, about the human consequences of elucidating the biology of consciousness. For some nonspecialists, consciousness and mind are virtually indistinguishable, and so are consciousness and conscience, and consciousness and soul, or consciousness and spirit. For them, maybe for you, mind, consciousness, conscience, soul, and spirit form one big region of strangeness that sets humans apart, that separates the mysterious from the explainable and the sacred from the profane. It should not be surprising to discover that the manner in which this sublime conflation of human properties is approached matters greatly to any sensible human being, and even that offense can be taken at seemingly dismissive accounts of its nature. Anyone who has faced death will know precisely what I am referring to, perhaps because the irreversibility of death focuses our thoughts sharply on the monumental scale of the human minded life. It should not take death, however, to make anyone sensitive to this issue. Life should be enough to make us approach the human mind with respect for its dignity and stature, and, almost paradoxically, with tenderness for its fragility.

Let me make something clear, however. Science helps us make distinctions among phenomena and science can now successfully distinguish among several components of the human mind. Consciousness and conscience are in fact distinguishable: consciousness pertains to the knowing of any object or action attributed to a self, while conscience pertains to the good or evil to be found in actions or objects. Consciousness and mind are also distinguishable: consciousness is the part of mind concerned with the apparent sense of self and knowing. There is more to mind than just consciousness and there can be mind without consciousness, as we discover in patients who have one but not the other.

In its progress, science proposes explanations for the phenomena it manages to distinguish. In the case of mind, it manages to explain parts of the big region of strangeness. It gleans *some* mechanisms behind *some* phenomena which contribute to the creation of the admirable human mind we so respect. Yet the admirable creation does not vanish just because we manage to explain some of the component mechanisms necessary for it to occur. The appearance *is* the reality—the human mind as we directly sense it. When we explain the mind, we get to keep that reality while we satisfy part of our curiosity regarding the sleight of hand behind the appearance.

Another issue I must make clear: solving the mystery of consciousness is not the same as solving all the mysteries of the mind. Consciousness is an indispensable ingredient of the creative human mind, but it is not all of human mind, and, as I see it, it is not the summit of mental complexity, either. The biological tricks that cause consciousness have powerful consequences, but I see consciousness as an intermediary rather than as the culmination of biological development. Ethics and the law, science and technology, the work of the muses and the milk of human kindness, those are my chosen summits for biology. Surely, we would have none of that without the wonders of consciousness at the source of each new achievement. Still, consciousness is a sunrise, not the midday sun, and a sunset even less. Understanding consciousness says little or nothing about the origins of the universe, the meaning of life, or the likely destiny of both. After solving the mystery of consciousness and making a dent on a few related mysteries of mind, assuming science achieves either, there is enough mystery left to last many a scientific lifetime, enough awe at nature to keep us modest for the foreseeable future. After considering how consciousness may be produced within the three pounds of flesh we call brain, we may revere life and respect human beings more, rather than less.

HIDE AND SEEK

Sometimes we use our minds not to discover facts but to hide them. We use part of the mind as a screen to prevent another part of it from sensing what goes on elsewhere. The screening is not necessarily intentional — we are not deliberate obfuscators all of the time — but deliberate or not, the screen does hide.

One of the things the screen hides most effectively is the body, our own body, by which I mean the ins of it, its interiors. Like a veil thrown over the skin to secure its modesty, but not too well, the screen partially removes from the mind the inner states of the body, those that constitute the flow of life as it wanders in the journey of each day.

The alleged vagueness, elusiveness, and intangibility of emotions and feelings is probably a symptom of this fact, an indication of how we cover the representation of our bodies, of how much mental imagery based on nonbody objects and events masks the reality of the body. Otherwise we would easily know that emotions and feelings are tangibly about the body. Sometimes we use our minds to hide a part of our beings from another part of our beings.

I could describe the hiding of the body as a distraction, but I would have to add that it is a very adaptive distraction. In most circumstances, rather than concentrating resources on our inner states, it is perhaps more advantageous to concentrate one's resources on the images that describe problems out in the world or on the premises of those problems or on the options for their solution and their possible outcomes. Yet this skewing of perspective relative to what is available in our minds has a cost. It tends to prevent us from sensing the possible origin and nature of what we call self. When the veil is lifted, however, at the scale of understanding permitted to the human mind, I believe we can sense the origin of the construct we call self in the representation of individual life.

Perhaps it was easier to get a more balanced perspective in earlier times when there was no veil, when the environments were relatively simple, long before electronic media and jet travel, long before the printed word, before the empire, and ahead of the city-state. It must have been easier to sense the life within, when the brain provided a lopsided view in the opposite direction, tilted toward the dominant representation of the internal states of the organism. If it ever was like that, perhaps at some magic brief time between Homer and Athens, lucky humans would have perceived in an instant that all of their amusing antics were about life and that underneath every image of the outside world, there stood the ongoing image of their living bodies. Or perhaps they would not have perceived as much because they lacked the frame of reference that current knowledge about biology provides us. Be that as it may, I suspect they were able to sense more about themselves than many of us, the unforewarned, are able to sense these days. I marvel at the ancient wisdom of referring to what we now call mind by the word *psyche* which was also used to denote breath and blood.

I suggest that the highly constrained ebb and flow of internal organism states, which is innately controlled by the brain and continuously signaled in the brain, constitutes the backdrop for the mind, and, more specifically, the foundation for the elusive entity we designate as self. I also suggest that those internal states—which occur naturally along a range whose poles are pain and pleasure, and are caused by either internal or external objects and events—become unwitting nonverbal signifiers of the goodness or badness of situations relative to the organism's inherent set of values. I suspect that in earlier stages of evolution these states—including all of those we classify as emotions—were entirely unknown to the organisms producing them. The states were regulatory and that was enough; they produced some advantageous actions, internally or externally, or they assisted indirectly the production of such actions by making them more propitious. But the organisms carrying out these complicated operations knew nothing of the existence of those operations and actions since they did not even know, in the proper sense of the word, of their own existence as individuals. True enough, organisms had a body and a brain, and brains had some representation of the body. Life was there, and the representation of life was there, too, but the potential and rightful owner of each individual life had no knowledge that life existed because nature had not invented an owner yet. There was being but not knowing. Consciousness had not begun.

Consciousness begins when brains acquire the power, the simple power I must add, of telling a story without words, the story that there is life ticking away in an organism, and that the states of the living organism, within body bounds, are continuously being altered by encounters with objects or events in its environment, or, for that matter, by thoughts and by internal adjustments of the life process. Consciousness emerges when this primordial story—the story of an object causally changing the state of the body—can be told using the universal nonverbal vocabulary of body signals. The apparent self emerges as the feeling of a feeling. When the story is first told, spontaneously, without it ever having been requested, and forevermore after that when the story is repeated, knowledge about what the organism is living through automatically emerges as the answer to a question never asked. From that moment on, we begin to know.

I suspect consciousness prevailed in evolution because knowing the feelings caused by emotions was so indispensable for the art of life, and because the art of life has been such a success in the history of nature. But I will not mind if you prefer to give my words a twist and just say that consciousness was invented so that we could know life. The wording is not scientifically correct, of course, but I like it.

PART II
Feeling and Knowing

Chapter Two

Emotion and Feeling

ONCE MORE WITH EMOTION

Without exception, men and women of all ages, of all cultures, of all levels of education, and of all walks of economic life have emotions, are mindful of the emotions of others, cultivate pastimes that manipulate their emotions, and govern their lives in no small part by the pursuit of one emotion, happiness, and the avoidance of unpleasant emotions. At first glance, there is nothing distinctively human about emotions since it is clear that so many nonhuman creatures have emotions in abundance; and yet there is something quite distinctive about the way in which emotions have become connected to the complex ideas, values, principles, and judgments that only humans can have, and in that connection lies our legitimate sense that human emotion is special. Human emotion is not just about sexual pleasures or fear of snakes. It is also about the horror of witnessing suffering and about the satisfaction of seeing justice served; about our delight at the sensuous smile of Jeanne Moreau or the thick beauty of words and ideas in Shakespeare's verse; about the world-weary voice of Dietrich Fischer-Dieskau singing Bach's *Ich habe genug* and the simultaneously earthly and otherworldly phrasings of Maria João Pires playing any Mozart, any Schubert; and about the harmony that Einstein sought in the structure of an equation. In fact, fine human emotion is even triggered by cheap music and cheap movies, the power of which should never be underestimated.

The human impact of all the above causes of emotion, refined and not so refined, and of all the shades of emotion they induce, subtle and not so subtle, depends on the feelings engendered by those emotions. It is through feelings, which are inwardly directed and private, that emotions, which are outwardly directed and public, begin their impact on the mind; but the full and lasting impact of feelings requires consciousness, because only along with the advent of a sense of self do feelings become known to the individual having them.

Some readers may be puzzled by the distinction between "feeling" and "knowing that we have a feeling." Doesn't the state of feeling imply, of necessity, that the feeler organism is fully conscious of the emotion and feeling that are unfolding? I am suggesting that it does not, that an organism may represent in neural and mental patterns the state that we conscious creatures call a feeling, without ever knowing that the feeling is taking place. This separation is difficult to envision, not only because the traditional meanings of the words block our view, but because we *tend* to be conscious of our feelings. There is, however, no evidence that we are conscious of *all* our feelings, and much to suggest that we are not. For example, we often realize quite suddenly, in a given situation, that we feel anxious or uncomfortable, pleased or relaxed, and it is apparent that the particular state of feeling we know then has not begun on the moment of knowing but rather sometime before. Neither the feeling state nor the emotion that led to it have been "in consciousness," and yet they have been unfolding as biological processes. These distinctions may sound artificial, at first glance, although my purpose is not to complicate something simple but rather to break down, in approachable parts, something that is quite complicated. For the purpose of investigating these phenomena, I separate three stages of processing along a continuum: *a state of emotion,* which can be triggered and executed nonconsciously; *a state of feeling,* which can be represented nonconsciously; and *a state of feeling made conscious,* i.e., known to the organism having both emotion and feeling. I believe these distinctions are helpful as we try to imagine the neural underpinnings of this chain of events in humans. Moreover, I suspect that some nonhuman creatures that exhibit emotions but are unlikely to have the sort of consciousness we have may well form the representations we call feelings without knowing they do so. Someone may suggest that perhaps we should have another word for "feelings that are not conscious," but there isn't one. The closest alternative is to explain what we mean.

In short, consciousness must be present if feelings are to influence the subject having them beyond the immediate here and now. The significance of this fact, that the ultimate consequences of human emotion and feeling pivot on consciousness, has not been properly appreciated (the strange history of research on emotion and feeling, addressed below, is possibly to blame for this neglect). Emotion was probably set in evolution before the dawn of consciousness and surfaces in each of us as a result of inducers we often do not recognize consciously; on the other hand, feelings perform their ultimate and longer-lasting effects in the theater of the conscious mind.

The powerful contrast between the covertly induced and outward posture of emotion and the inwardly directed and ultimately known status of human feeling provided me with an invaluable perspective for reflection on the biology of consciousness. And there are other bridges between emotion and consciousness. In this book, I propose that, just like emotion, consciousness is aimed at the organism's survival, and that, just like emotion, consciousness is rooted in the representation of the body. I also call attention to an intriguing neurological fact: when consciousness is suspended, from core consciousness on up, emotion is usually suspended as well, suggesting that although emotion and consciousness are different phenomena, their underpinnings may be connected. For all these reasons, it is important to discuss the varied features of emotion before we begin addressing consciousness directly. But first, before I outline the results of that reflection, I propose an aside on the strange history of the science of emotion, because that history may help explain why consciousness has not been approached from the perspective I am adopting here.

A Historical Aside

Given the magnitude of the matters to which emotion and feeling have been attached, one would have expected both philosophy and the sciences of mind and brain to have embraced their study. Surprisingly, that is only happening now. Philosophy, notwithstanding David Hume and the tradition that originates with him, has not trusted emotion and has largely relegated it to the dismissible realms of animal and flesh. For a time, science fared better, but then it, too, missed its opportunity.

By the end of the nineteenth century Charles Darwin, William James, and Sigmund Freud had written extensively on different aspects of emotion and given emotion a privileged place in scientific discourse. Yet, throughout the twentieth century and until quite recently, both neuroscience and cognitive science gave emotion a very cold shoulder. Darwin had conducted an extensive study of the expression of emotion in different cultures and different species, and though he thought of human emotions as vestiges from previous stages of evolution, he respected the importance of the phenomenon. William James had seen through the problem with his characteristic clarity and produced an account that, in spite of its incompleteness, remains a cornerstone. As for Freud, he had gleaned the pathological potential of disturbed emotions and announced their importance in no uncertain terms.

Darwin, James, and Freud were, of necessity, somewhat vague about the brain aspect of their ideas, but one of their contemporaries,

Hughlings Jackson, was more precise. He took the first step toward a possible neuroanatomy of emotion and suggested that the right cerebral hemisphere of humans was probably dominant for emotion, much as the left was dominant for language.

There would have been good reason to expect that, as the new century started, the expanding brain sciences would make emotion part of their agenda and solve its questions. But that development never came to pass. Worse than that, Darwin's work on the emotions vanished from sight, James's proposal was attacked unfairly and dismissed summarily, and Freud's influence went elsewhere. Throughout most of the twentieth century, emotion was not trusted in the laboratory. Emotion was too subjective, it was said. Emotion was too elusive and vague. Emotion was at the opposite end from reason, easily the finest human ability, and reason was presumed to be entirely independent from emotion. This was a perverse twist on the Romantic view of humanity. Romantics placed emotion in the body and reason in the brain. Twentieth-century science left out the body, moved emotion back into the brain, but relegated it to the lower neural strata associated with ancestors who no one worshiped. In the end, not only was emotion not rational, even studying it was probably not rational.

There are curious parallels to the scientific neglect of emotion during the twentieth century. One of those parallels is the lack of an *evolutionary perspective* in the study of brain and mind. It is perhaps an exaggeration to say that neuroscience and cognitive science have proceeded as if Darwin never existed, but it certainly seemed so until the last decade. Aspects of brain and mind have been discussed as if designed recently, as needed, to produce a certain effect—a bit like the installation of antilock brakes in a proper new car—without any regard for the possible antecedents of mental and brain devices. Of late the situation is changing remarkably.

Another parallel concerns the disregard for the notion of homeostasis. *Homeostasis* refers to the coordinated and largely automated physiological reactions required to maintain steady internal states in a living organism. Homeostasis describes the automatic regulation of temperature, oxygen concentration, or pH in your body. Numerous scientists have been preoccupied with understanding the neurophysiology of homeostasis, with making sense of the neuroanatomy and the neurochemistry of the autonomic nervous system (the part of the nervous system most directly involved in homeostasis), and with elucidating the interrelations among the endocrine, immune, and nervous systems, whose ensemble work produces homeostasis. But the scientific progress made in those areas had little influence on the prevailing views of how mind or brain worked. Curiously enough, emotions are part and parcel of the regulation we call homeostasis. It is senseless to discuss them without understanding that aspect of living organisms and vice versa. In this book, I propose that homeostasis is a key to the biology of consciousness (see chapter 5).

A third parallel is the noticeable absence of a notion of *organism* in cognitive science and neuroscience. The mind remained linked to the brain in a somewhat equivocal relationship, and the brain remained consistently separated from the body rather than being seen as part of a complex living organism. The notion of an integrated organism— the idea of an ensemble made up of a body proper and a nervous system—was available in the work of thinkers such as Ludwig von Bertalanffy, Kurt Goldstein, and Paul Weiss but had little impact in shaping the standard conceptions of mind and brain.[1]

To be sure, there are exceptions in this broad panorama. For instance, Gerald Edelman's theoretical proposals on the neural basis of the mind are informed by evolutionary thinking and acknowledge homeostatic regulation; and my somatic-marker hypothesis is grounded on notions of evolution, homeostatic regulation, and organism.[2] But the theoretical assumptions according to which cognitive science and neuroscience have been conducted have not made much use of organismic and evolutionary perspectives.

In recent years both neuroscience and cognitive neuroscience have finally endorsed emotion. A new generation of scientists is now making emotion its elected topic.[3] Moreover, the presumed opposition between emotion and reason is no longer accepted without question.

For example, work from my laboratory has shown that emotion is integral to the processes of reasoning and decision making, for worse and for better.[4] This may sound a bit counterintuitive, at first, but there is evidence to support it. The findings come from the study of several individuals who were entirely rational in the way they ran their lives up to the time when, as a result of neurological damage in specific sites of their brains, they lost a certain class of emotions and, in a momentous parallel development, lost their ability to make rational decisions. Those individuals can still use the instruments of their rationality and can still call up the knowledge of the world around them. Their ability to tackle the logic of a problem remains intact. Nonetheless, many of their personal and social decisions are irrational, more often disadvantageous to their self and to others than not. I have suggested that the delicate mechanism of reasoning is no longer affected, nonconsciously and on occasion even consciously, by signals hailing from the neural machinery that underlies emotion.

This hypothesis is known as the somatic-marker hypothesis, and the patients who led me to propose it had damage to selected areas in the prefrontal region, especially in the ventral and medial sectors, and in the right parietal regions. Whether because of a stroke or head injury or a tumor which required surgical resection, damage in those regions was consistently associated with the appearance of the clinical pattern I described above, i.e., a disturbance of the ability to decide advantageously in situations involving risk and conflict and a selective reduction of the ability to resonate emotionally in precisely those same situations, while preserving the remainder of their emotional abilities. Prior to the onset of their brain damage, the individuals so affected had shown no such impairments. Family and friends could sense a "before" and an "after," dating to the time of neurologic injury.

These findings suggest that selective reduction of emotion is at least as prejudicial for rationality as excessive emotion. It certainly does not seem true that reason stands to gain from operating without the leverage of emotion. On the contrary, emotion probably assists reasoning, especially when it comes to personal and social matters involving risk and conflict. I suggested that certain levels of emotion processing probably point us to the sector of the decision-making space where our reason can operate most efficiently. I did *not* suggest, however, that emotions are a substitute for reason or that emotions decide for us. It is obvious that emotional upheavals can lead to irrational decisions. The neurological evidence simply suggests that selective absence of emotion is a problem. Well-targeted and well-deployed emotion seems to be a support system without which the edifice of reason cannot operate properly. These results and their interpretation called into question the idea of dismissing emotion as a luxury or a nuisance or a mere evolutionary vestige. They also made it possible to view emotion as an embodiment of the logic of survival.[5]

THE BRAIN KNOWS MORE THAN THE CONSCIOUS MIND REVEALS

Emotions and feelings of emotions, respectively, are the beginning and the end of a progression, but the relative publicness of emotions and the complete privacy of the ensuing feelings indicate that the mechanisms along the continuum are quite different. Honoring a distinction between emotion and feeling is helpful if we are to investigate those mechanisms thoroughly. I have proposed that the term *feeling* should be reserved for the private, mental experience of an emotion, while the term *emotion* should be used to designate the collection of responses, many of which are publicly observable. In practical terms this means that you cannot observe a feeling in someone else although you can observe a feeling in yourself when, as a conscious being, you perceive your own emotional states. Likewise no one can observe your own feelings, but some aspects of the emotions that give rise to your feelings will be patently observable to others. Moreover, for the sake of my argument, the basic mechanisms underlying emotion do not require consciousness, even if they eventually use it: you can initiate the cascade of processes that lead to an emotional display without being conscious of the inducer of the emotion let alone the intermediate steps leading to it. In effect, even the occurrence of a feeling in the limited time window of the here and now is conceivable without the organism actually *knowing* of its occurrence. To be sure, at this point in evolution and at this moment of our adult lives, emotions occur in a setting of consciousness: We can feel our emotions consistently and we know we feel them. The fabric of our minds and of our behavior is woven around continuous cycles of emotions followed by feelings that become known and beget new emotions, a running polyphony that underscores and punctuates specific thoughts in our minds and actions in our behavior. But although emotion and feeling are now part of a functional continuum, it is helpful to distinguish the steps along that continuum if we are to study their biological underpinnings with any degree of success. Besides, as suggested earlier, it is possible that feelings are poised at the very threshold that separates being from knowing and thus have a privileged connection to consciousness.[6]

WHY AM I so confident that the biological machinery underlying emotion is not dependent on consciousness? After all, in our daily experience, we often seem to know the circumstances leading to an emotion. But knowing often is not the same as knowing always. There is good evidence in favor of the covert nature of emotion induction, and I will illustrate the point with some experimental results from my laboratory.

David, who has one of the most severe defects in learning and memory ever recorded, cannot learn any new fact at all. For instance, he cannot learn any new physical appearance or sound or place or word. As a consequence he cannot learn to recognize any new person, from the face, from the voice, or from the name, nor can he remember anything whatsoever regarding where he has met a certain person or the events that transpired between him and that person. David's problem is caused by extensive damage to both temporal lobes, which includes damage·to a region known as the hippocampus (whose integrity is necessary to create memories for new facts) and the region known as the amygdala (a subcortical grouping of nuclei concerned with emotion that I will mention in the pages ahead).

Many years ago I heard that David seemed to manifest, in his day-to-day life, consistent preferences and avoidances for certain persons. For instance, in the facility where he has lived for most of the past twenty years, there were specific people whom he would frequently choose to go to if he wanted a cigarette or a cup of coffee, and there were certain people to whom he would never go. The consistency of these behaviors was most intriguing, considering that David could not recognize any of those individuals at all; considering that he had no idea whether he had ever seen any of them; and considering that he could not produce the name of any of them or point to any of them given the name. Could this intriguing story be more than a curious anecdote? I decided to check it out and put it to empirical test. In order to do so, I collaborated with my colleague Daniel Tranel to design an experiment which has become known in our laboratory as the good-guy/bad-guy experiment.[7]

Over a period of a week, we were able to engage David, under entirely controlled circumstances, in three distinct types of human interaction. One type of interaction was with someone who was extremely pleasant and welcoming and who always rewarded David whether he requested something or not (this was the good guy). Another interaction involved somebody who was emotionally neutral and who engaged David in activities that were neither pleasant nor unpleasant (this was the neutral guy). A third type of interaction involved an individual whose manner was brusque, who would say no to any request, and who engaged David in a very tedious psychological task designed to bring boredom to a saint (this was the bad guy). The task was the delayed-nonmatching-to-sample task, which was invented to investigate memory in monkeys and is probably a delight if you have the mind of a monkey.

The staging of these different situations was arranged throughout five consecutive days, in random order, but always for a specified interval of time so that the overall exposure to the good, to the bad, and to the indifferent would be properly measured and compared. The elaborate staging of this dance required varied rooms and several assistants, who were not the same, by the way, as the good, bad, and neutral guys.

After all the encounters were allowed to sink in, we asked David to participate in two distinct tasks. In one task David was asked to look at sets of four photographs that included the face of one of the three individuals in the experiment, and then asked, "Whom would you go to if you needed help?" and, for further clarification, "Who do you think is your friend in this group?"

David behaved in a most spectacular manner. When the individual who had been positive to him was part of the set of four, David chose the good guy over 80 percent of the time, indicating that his choice was clearly not random—chance alone would have made David pick each of the four 25 percent of the time. The neutral individual was chosen with a probability no greater than chance. And the bad guy was almost never chosen, again something that violated chance behavior.

In a second task, David was asked to look at the faces of the three individuals and tell us what he knew about them. As usual, for him, nothing came to mind. David could not remember ever encountering them and had no recollection of any instance in which he had interacted with them. Needless to say, he could not name any of the individuals, he could not point to any of them given the name, and he had no idea of what we were talking about when we asked him about the events of the previous week. But when he was asked who, among the three, was his friend, he consistently chose the good guy.

The results show that the anecdote was well worth investigating. To be sure, there was nothing in David's conscious mind that gave him an overt reason to choose the good guy correctly and reject the bad one correctly. He did not know why he chose one or rejected the other; he just did. The nonconscious preference he manifested, however, is probably related to the emotions that were induced in him during the experiment, as well as to the nonconscious reinduction of some part of those emotions at the time he was being tested. David had not learned new knowledge of the type that can be deployed in one's mind in the form of an image. But something stayed in his brain and that something could produce results in nonimage form: in the form of actions and behavior. David's brain could generate actions commensurate with the emotional value of the original encounters, as caused by reward or lack thereof. To make this idea clear, let me describe an observation I made on one occasion during the exposure sessions in the good-guy/bad-guy experiment.

David was being brought to a bad-guy encounter and as he turned into the hallway and saw the bad guy awaiting him, a few feet away, he flinched, stopped for an instant, and only then allowed himself to be led gently to the examining room. I picked up on this and immediately asked him if anything was the matter, if there was anything I could do for him. But, true to form, he told me that, no, everything was all right—after all, nothing came to his mind, except, perhaps, an isolated sense of emotion without a cause behind that emotion. I have no doubt that the sight of the bad guy induced a brief emotional response and a brief here-and-now feeling. However, in the absence of an appropriately related set of images that would explain to him the cause of the reaction, the effect remained isolated, disconnected, and thus unmotivated.[8]

I also have little doubt that were we to have carried out this task for weeks in a row rather than for one single week, David would have harnessed such negative and positive responses to produce the behavior that suited his organism best, i.e., prefer the good guy consistently and avoid the bad guy. But I am not suggesting that *he* himself would have chosen to do so deliberately, but rather that his *organism,* given its available design and dispositions, would have homed in on such behavior. He would have developed a tropism for the good guy as well as an antitropism for the bad guy, in much the same manner he had developed such preferences in the real-life setting.

The situation just described allows us to make some other points. First, David's core consciousness is intact, an issue that we shall revisit in the next chapter. Second, while in the setting of the good-guy/bad-guy experiment, David's emotions were induced nonconsciously, in other settings he engages emotions knowingly. When he does not have to depend on new memory, he senses that he is happy because he is tasting a favorite food or watching a pleasant scene. Third, given the remarkable destruction of several cortical and subcortical emotion-related regions of his brain, e.g., ventromedial prefrontal cortices, basal forebrain, amygdalae, it is apparent that those territories are not indispensable for either emotion or consciousness. We may also keep in mind, for future reference, that certain structures of David's brain remain intact: all of the brain stem; the hypothalamus; the thalamus; most of the cingulate cortices; and virtually all sensory and motor structures.

Let me close these comments by saying that the bad guy in our experiment was a young, pleasant, and beautiful woman neuropsychologist. We had designed the experiment in this way, having her play against type, since we wanted to determine the extent to which David's manifest preference for the company of young and beautiful women would countervail the contrariness of her behavior and the fact that she was the purveyor of the boring task (David does have an eye for the girls; I caught him once caressing Patricia Churchland's arm and remarking, "You are so soft . . ."). Well, as you can see, our benign bit of perverse planning paid off. No amount of natural beauty could have compensated for the negative emotion induced by the bad guy's manner and by the poor entertainment her task provided.

WE DO NOT need to be conscious of the inducer of an emotion and often are not, and we cannot control emotions willfully. You may find yourself in a sad or happy state, and yet you may be at a loss as to why you are in that particular state now. A careful search may disclose possible causes, and one cause or another may be more likely, but often you cannot be certain. The actual cause may have been the image of an event, an image that had the potential to be conscious but just was not because you did not attend to it while you were attending to another. Or it may have been no image at all, but rather a transient change in the chemical profile of your internal milieu, brought about by factors as diverse as your state of health, diet, weather, hormonal cycle, how much or how little you exercised that day, or even how much you had been worrying about a certain matter. The change would be substantial enough to engender some responses and alter your body state, but it would not be imageable in the sense that a person or a relationship are imageable, i.e., it would not produce a sensory pattern of which you would ever become aware in your mind. In other words, the representations which induce emotions and lead to subsequent feelings need not be attended, regardless of whether they signify something external to the organism or something recalled internally. Representations of either the exterior or the interior can occur underneath conscious survey and still induce emotional responses. Emotions can be induced in a nonconscious manner and thus appear to the conscious self as seemingly unmotivated.

We can control, in part, whether a would-be inducer image should be allowed to remain as a target of our thoughts. (If you were raised Catholic you know precisely what I mean, and likewise, if you have been around the Actors Studio.) We may not succeed at the task, but the job of removing or maintaining the inducer certainly occurs in consciousness. We can also control, in part, the expression of some emotions—suppress our anger, mask our sadness—but most of us are not very good at it and that is one reason why we pay a lot to see good actors who are skilled at controlling the expression of their emotions (and why we may lose a lot of money playing poker). Once a particular sensory representation is formed, however, whether or not it is actually part of our conscious thought flow, we do not have much to say on the mechanism of inducing an emotion. If the psychological and physiological context is right, an emotion will ensue. The nonconscious triggering of emotions also explains why they are not easy to mimic voluntarily. As I explained in *Descartes' Error*, a spontaneous smile that comes from genuine delight or the spontaneous sobbing that is caused by grief are executed by brain structures located deep in the brain stem under the control of the cingulate region. We have no means of exerting direct voluntary control over the neural processes in those regions. Casual voluntary mimicking of expressions of emotion is easily detected as fake — something always fails, whether in the configuration of the facial muscles or in the tone of voice. The result of this state of affairs is that in most of us who are not actors, emotions are a fairly good index of how conducive the environment is to our well-being, or, at least, how conducive it seems to our minds.

We are about as effective at stopping an emotion as we are at preventing a sneeze. We can try to prevent the expression of an emotion, and we may succeed in part but not in full. Some of us, under the appropriate cultural influence, get to be quite good at it, but in essence what we achieve is the ability to disguise some of the external manifestations of emotion without ever being able to block the automated changes that occur in the viscera and internal milieu. Think of the last time you were moved in public and tried to disguise it. You might have gotten away with it if you were watching a movie out there in the dark, alone with Gloria Swanson, but not if you were delivering the eulogy for a dead friend: your voice would have given you away. Someone once told me that the idea of feelings occurring after emotion could not be correct since it is possible to suppress emotions and still have feelings. But that is not true, of course, beyond the partial suppression of facial expressions. We can educate our emotions but not suppress them entirely, and the feelings we have inside are a testimony to our lack of success.

An Aside on Controlling the Uncontrollable

One partial exception to the extremely limited control we have over the internal milieu and viscera concerns respiratory control, over which we need to exert some voluntary action, because autonomic respiration and voluntary vocalization for speech and singing use the same instrument. You can learn to swim underwater, holding your breath for longer and longer periods, but there are limits beyond which no Olympic champion can go and remain alive. Opera singers face a similar barrier: what tenor wouldn't love to hold on to the high C for just a while longer and irritate the soprano? But no amount of laryngeal and diaphragmatic training will allow tenor or soprano to transpose the barrier. Indirect control of blood pressure and heart rate by procedures such as biofeedback are also partial exceptions. As a rule, however, voluntary control over autonomic function is modest.

I can report one dramatic exception, however. Some years ago the brilliant pianist Maria João Pires told us the following story: When she plays, under the perfect control of her will, she can either reduce or allow the flow of emotion to her body. My wife, Hanna, and I thought this was a wonderfully romantic idea, but Maria João insisted that she could do it and we resisted believing it. Eventually, the stage for the empirical moment of truth was set in our laboratory. Maria João was wired to the complicated psychophysiological equipment while she listened to short musical pieces of our selection in two conditions: emotion allowed, or emotion voluntarily inhibited. Her Chopin *Nocturnes* had just been released, and we used some of hers and some of Daniel Barenboim's as stimuli. In the condition of "emotion allowed," her skin conductance record was full of peaks and valleys, linked intriguingly to varied passages in the pieces. Then, in the condition of "emotion reduced," the unbelievable did, in fact, happen. She could virtually flatten her skin-conductance graph at will and change her heart rate, to boot. Behaviorally, she changed as well. The profile of background emotions was rearranged, and some of the specific emotive behaviors were eliminated, e.g., there was less movement of the head and facial musculature. When our colleague Antoine Bechara, in complete disbelief, repeated the whole experiment, wondering if this might be an artifact of habituation, she did it again. So there are some exceptions to be found after all, perhaps more so in those whose lifework consists of creating magic through emotion.

WHAT ARE EMOTIONS?

The mention of the word *emotion* usually calls to mind one of the six so-called *primary* or *universal emotions:* happiness, sadness, fear, anger, surprise, or disgust. Thinking about the primary emotions makes the discussion of the problem easier, but it is important to note that there are numerous other behaviors to which the label "emotion" has been attached. They include so-called *secondary* or *social emotions,* such as embarrassment, jealousy, guilt, or pride; and what I call *background emotions,* such as well-being or malaise, calm or tension. The label emotion has also been attached to drives and motivations and to the states of pain and pleasure.[9]

A shared biological core underlies all these phenomena, and it can be outlined as follows:

1. Emotions are complicated collections of chemical and neural responses, forming a pattern; all emotions have some kind of regulatory role to play, leading in one way or another to the creation of circumstances advantageous to the organism exhibiting the phenomenon; emotions are *about* the life of an organism, its body to be precise, and their role is to assist the organism in maintaining life.

2. Notwithstanding the reality that learning and culture alter the expression of emotions and give emotions new meanings, emotions are biologically determined processes, depending on innately set brain devices, laid down by a long evolutionary history.

3. The devices which produce emotions occupy a fairly restricted ensemble of subcortical regions, beginning at the level of the brain stem and moving up to the higher brain; the devices are part of a set of structures that both regulate and represent body states, which will be discussed in chapter 5.

4. All the devices can be engaged automatically, without conscious deliberation; the considerable amount of individual variation and the fact that culture plays a role in shaping some inducers does not deny the fundamental stereotypicity, automaticity, and regulatory purpose of the emotions.

5. All emotions use the body as their theater (internal milieu, visceral, vestibular and musculoskeletal systems), but emotions also affect the mode of operation of numerous brain circuits:

the variety of the emotional responses is responsible for profound changes in both the body landscape and the brain landscape. The collection of these changes constitutes the substrate
for the neural patterns which eventually become feelings of
emotion.

A special word about background emotions is needed, at this point,
because the label and the concept are not a part of traditional discussions on emotion. When we sense that a person is "tense" or "edgy,"
"discouraged" or "enthusiastic," "down" or "cheerful," without a
single word having been spoken to translate any of those possible
states, we are detecting background emotions. We detect background
emotions by subtle details of body posture, speed and contour of
movements, minimal changes in the amount and speed of eye movements, and in the degree of contraction of facial muscles.

The inducers of background emotions are usually internal. The
processes of regulating life itself can cause background emotions but
so can continued processes of mental conflict, overt or covert, as they
lead to sustained satisfaction or inhibition of drives and motivations.
For example, background emotions can be caused by prolonged physical effort—from the "high" that follows jogging to the "low" of un-
interesting, nonrhythmical physical labor—and by brooding over a
decision that you find difficult to make—one of the reasons behind
Prince Hamlet's dispirited existence—or by savoring the prospect of
some wonderful pleasure that may await you. In short, certain conditions of internal state engendered by ongoing physiological processes
or by the organism's interactions with the environment or both cause
responses which constitute background emotions. Those emotions
allow us to have, among others, the background feelings of tension or
relaxation, of fatigue or energy, of well-being or malaise, of anticipation or dread.[10]

In background emotions, the constitutive responses are closer to
the inner core of life, and their target is more internal than external.
Profiles of the internal milieu and viscera play the lead part in back-

ground emotions. But although background emotions do not use the differentiated repertoire of explicit facial expressions that easily define primary and social emotions, they are also richly expressed in musculoskeletal changes, for instance, in subtle body posture and overall shaping of body movement.[11]

In my experience, background emotions are brave survivors of neurological disease. For instance, patients with ventromedial frontal damage retain them, as do patients with amygdala damage. Intriguingly, as you will discover in the next chapter, background emotions are usually compromised when the basic level of consciousness, core consciousness, is compromised as well.

THE BIOLOGICAL FUNCTION OF EMOTIONS

Although the precise composition and dynamics of the emotional responses are shaped in each individual by a unique development and environment, the evidence suggests that most, if not all, emotional responses are the result of a long history of evolutionary fine-tuning. Emotions are part of the bioregulatory devices with which we come equipped to survive. That is why Darwin was able to catalog the emotional expressions of so many species and find consistency in those expressions, and that is why, in different parts of the world and across different cultures, emotions are so easily recognized. Surely enough, there are variable expressions and there are variations in the precise configuration of stimuli that can induce an emotion across cultures and among individuals. But the thing to marvel at, as you fly high above the planet, is the similarity, not the difference. It is that similarity, incidentally, that makes cross-cultural relations possible and that allows for art and literature, music and film, to cross frontiers. This view has been given immeasurable support by the work of Paul Ekman.[12]

The biological function of emotions is twofold. The first function is the production of a specific reaction to the inducing situation. In an animal, for instance, the reaction may be to run or to become immobile or to beat the hell out of the enemy or to engage in pleasurable

behavior. In humans, the reactions are essentially the same, tempered, one hopes, by higher reason and wisdom. The second biological function of emotion is the regulation of the internal state of the
organism such that it can be prepared for the specific reaction. For example, providing increased blood flow to arteries in the legs so that
muscles receive extra oxygen and glucose, in the case of a flight reaction, or changing heart and breathing rhythms, in the case of freezing
on the spot. In either case, and in other situations, the plan is exquisite and the execution is most reliable. In short, for certain classes of
clearly dangerous or clearly valuable stimuli in the internal or external environment, evolution has assembled a matching answer in the
form of emotion. This is why, in spite of the infinite variations to be
found across cultures, among individuals, and over the course of a life
span, we can predict with some success that certain stimuli will produce certain emotions. (This is why you can say to a colleague, "Go
tell her that; she will be so happy to hear it.")

In other words, the biological "purpose" of the emotions is clear,
and emotions are not a dispensable luxury. Emotions are curious
adaptations that are part and parcel of the machinery with which organisms regulate survival. Old as emotions are in evolution, they are
a fairly high-level component of the mechanisms of life regulation.
You should imagine this component as sandwiched between the basic
survival kit (e.g., regulation of metabolism; simple reflexes; motivations; biology of pain and pleasure) and the devices of high reason,
but still very much a part of the hierarchy of life-regulation devices.
For less-complicated species than humans, and for absentminded humans as well, emotions actually produce quite reasonable behaviors
from the point of view of survival.

At their most basic, emotions are part of homeostatic regulation
and are poised to avoid the loss of integrity that is a harbinger of death
or death itself, as well as to endorse a source of energy, shelter, or sex.
And as a result of powerful learning mechanisms such as conditioning, emotions of all shades eventually help connect homeostatic
regulation and survival "values" to numerous events and objects in

our autobiographical experience. Emotions are inseparable from the idea of reward or punishment, of pleasure or pain, of approach or withdrawal, of personal advantage and disadvantage. Inevitably, emotions are inseparable from the idea of good and evil.

Table 2.1. Levels of Life Regulation

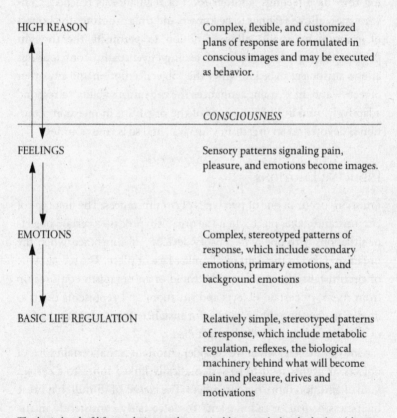

HIGH REASON	Complex, flexible, and customized plans of response are formulated in conscious images and may be executed as behavior.
CONSCIOUSNESS	
FEELINGS	Sensory patterns signaling pain, pleasure, and emotions become images.
EMOTIONS	Complex, stereotyped patterns of response, which include secondary emotions, primary emotions, and background emotions
BASIC LIFE REGULATION	Relatively simple, stereotyped patterns of response, which include metabolic regulation, reflexes, the biological machinery behind what will become pain and pleasure, drives and motivations

The basic level of life regulation—the survival kit—includes the biological states that can be consciously perceived as drives and motivations and as states of pain and pleasure. Emotions are at a higher, more complex level. The dual arrows indicate upward or downward causation. For instance, pain can induce emotions, and some emotions can include a state of pain.

One might wonder about the relevance of discussing the biological role of the emotions in a text devoted to the matter of consciousness. The relevance should become clear now. Emotions automatically provide organisms with survival-oriented behaviors. In organisms equipped to sense emotions, that is, to have feelings, emotions also have an impact on the mind, as they occur, in the here and now. But in organisms equipped with consciousness, that is, capable of knowing they have feelings, another level of regulation is reached. Consciousness allows feelings to be known and thus promotes the impact of emotion internally, allows emotion to permeate the thought process through the agency of feeling. Eventually, consciousness allows any object to be known—the "object" emotion and any other object—and, in so doing, enhances the organism's ability to respond adaptively, mindful of the needs of the organism in question. Emotion is devoted to an organism's survival, and so is consciousness.

INDUCING EMOTIONS

Emotions occur in one of two types of circumstances. The first type of circumstance takes place when the organism processes certain objects or situations with one of its sensory devices—for instance, when the organism takes in the sight of a familiar face or place. The second type of circumstance occurs when the mind of an organism conjures up from memory certain objects and situations and represents them as images in the thought process—for instance, remembering the face of a friend and the fact she has just died.

One obvious fact when we consider emotions is that certain sorts of objects or events tend to be systematically linked more to a certain kind of emotion more than to others. The classes of stimuli that cause happiness or fear or sadness tend to do so fairly consistently in the same individual and in individuals who share the same social and cultural background. In spite of all the possible individual variations in the expression of an emotion and in spite of the fact that we can have mixed emotions, there is a rough correspondence between classes of emotion inducers and the resulting emotional state. Throughout evolution, organisms have acquired the means to respond to certain stimuli—particularly those that are potentially useful or potentially dangerous from the point of view of survival—with the collection of responses which we currently call an emotion.

But a word of caution is needed here. I really mean what I say when I talk about *ranges of stimuli* that constitute inducers for certain *classes of emotion.* I am allowing for a considerable variation in the type of stimuli that can induce an emotion—both across individuals and across cultures—and I am calling attention to the fact that regardless of the degree of biological presetting of the emotional machinery, development and culture have much to say regarding the final product. In all probability, development and culture superpose the following influences on the preset devices: first, they shape what constitutes an adequate inducer of a given emotion; second, they shape some aspects of the expression of emotion; and third, they shape the cognition and behavior which follows the deployment of an emotion.[13]

It is also important to note that while the biological machinery for emotions is largely preset, the inducers are not part of the machinery, they are external to it. The stimuli that cause emotions are by no means confined to those that helped shape our emotional brain during evolution and which can induce emotions in our brains from early in life. As they develop and interact, organisms gain factual and emotional experience with different objects and situations in the environment and thus have an opportunity to associate many objects and situations which would have been emotionally neutral with the objects and situations that are naturally prescribed to cause emotions. A form of learning known as conditioning is one way of achieving this association. A new house of a shape similar to the house in which you lived a blissful childhood may make you feel well even if nothing especially good has yet happened to you in it. Likewise, the face of a wonderful, unknown person that so resembles that of someone associated with some horrible event may cause you discomfort or irritation. You may never come to know why. Nature did not prescribe those responses, but it surely helped you acquire them. Incidentally, superstitions are born this way. There is something Orwellian about the distribution of emotions in our world: All objects can get some emotional attachment, but some objects get far more than others. Our primary biological design skews our secondary acquisitions relative to the world around us.

The consequence of extending emotional value to objects that were not biologically prescribed to be emotionally laden is that the range of stimuli that can potentially induce emotions is infinite. In one way or another, most objects and situations lead to some emotional reaction, although some far more so than others. The emotional reaction may be weak or strong—and fortunately for us it is weak more often than not—but it is there nonetheless. Emotion and the biological machinery underlying it are the obligate accompaniment of behavior, conscious or not. Some level of emoting is the obligate accompaniment of thinking about oneself or about one's surroundings.

The pervasiveness of emotion in our development and subsequently in our everyday experience connects virtually every object or situation in our experience, by virtue of conditioning, to the fundamental values of homeostatic regulation: reward and punishment; pleasure or pain; approach or withdrawal; personal advantage or disadvantage; and, inevitably, good (in the sense of survival) or evil (in the sense of death). Whether we like it or not, this is the *natural* human condition. But when consciousness is available, feelings have their maximum impact, and individuals are also able to reflect and to plan. They have a means to control the pervasive tyranny of emotion: it is called reason. Ironically, of course, the engines of reason still require emotion, which means that the controlling power of reason is often modest.

Another important consequence of the pervasiveness of emotions is that virtually every image, actually perceived or recalled, is accompanied by some reaction from the apparatus of emotion. We will consider the importance of this fact when we discuss the mechanisms for the birth of consciousness in chapter 6.

Let me close this comment on inducers of emotions with a reminder of a tricky aspect of the induction process. So far, I have referred to direct inducers—thunder, snakes, happy memories. But emotions can be induced indirectly, and the inducer can produce its result in a somewhat negative fashion, by blocking the progress of an ongoing emotion. Here is an example. When, in the presence of a source of food or sex, an animal develops approach behavior and exhibits features of the emotion happiness, blocking its way and preventing it from achieving its goals will cause frustration and even anger, a very different emotion from happiness. The inducer of the anger is not the prospect of food or sex but rather the thwarting of the behavior that was leading the animal to the good prospect. Another example would be the sudden suspension of a situation of punishment—for instance, sustained pain—which would induce well-being and happiness. The purifying (cathartic) effect that all good tragedies should have, according to Aristotle, is based on the sudden suspension of a steadily induced state of fear and pity. Long after Aristotle, Alfred Hitchcock built a brilliant career on this simple biological arrangement, and Hollywood has never stopped banking on it. Whether we like it or not, we feel very comfortable after Janet Leigh stops screaming in the shower and lies quietly on the bathtub floor. As far as emotion goes, there is not much escape in the setup that nature prepared for us. We get it coming and we get it going.

The Mechanics of Emotion

From experience, you know that the responses that make up emotions are most varied. Some responses are easily apparent in yourself and in others. Think of the muscles in the face adopting the configurations that are typical of joy or sorrow or anger; or of the skin blanching as a reaction to bad news or flushing in a situation of embarrassment; or consider the body postures that signify joy, defiance, sadness, or discouragement; or the sweaty and clammy hands of apprehension; the racing heart associated with pride; or the slowing, near-stillness of the heart in terror.

Other responses are hidden from sight but no less important, such as the myriad changes that occur in organs other than blood vessels, skin, and heart. One example is the secretion of hormones such as cortisol that change the chemical profile of the internal milieu; or the secretion of peptides, such as β-endorphin or oxytocin, that alter the operation of several brain circuits. Another is the release of neurotransmitters, such as the monoamines, norepinephrine, serotonin, and dopamine. During emotions, neurons located in the hypothalamus, basal forebrain, and brain stem release those chemical substances in several regions of the brain up above and, by so doing, temporarily transform the mode of working for many neural circuits. Typical consequences of the increase or decrease of release of such transmitters include the sense we have of the mind processes speeding up or slowing down, not to mention the sense of pleasantness or unpleasantness that pervades mental experience. Such sensing is part of our feeling of an emotion.

Different emotions are produced by different brain systems. In the very same way that you can tell the difference between a facial expression of anger and a facial expression of joy, in the very same way in which you can feel the difference between sadness or happiness in your flesh, neuroscience is beginning to show us how different brain systems work to produce, say, anger or sadness or happiness.

The study of patients with neurological diseases and focal brain damage has yielded some of the most revealing results in this area, but these investigations are now being complemented by functional neuroimaging of individuals without neurological disease. I should note that the work with human subjects also permits a rich dialogue with investigators who are approaching some of these same problems in animals, another welcome novelty in this area of research.

The essence of the available findings can be summarized as follows. First, the brain induces emotions from a remarkably small number of brain sites. Most of them are located below the cerebral cortex and are known as subcortical. The main subcortical sites are in the brain-stem region, hypothalamus, and basal forebrain. One example is the region known as periaqueductal gray (PAG), which is a major coordinator of emotional responses. The PAG acts via motor nuclei of the reticular formation and via the nuclei of cranial nerves, such as the nuclei of the vagus nerve.[14] Another important subcortical site is the amygdala. The induction sites in the cerebral cortex, the cortical sites, include sectors of the anterior cingulate region and of the ventromedial prefrontal region.

Second, these sites are involved in processing different emotions to varying degrees. We have recently shown, using PET imaging, that the induction and experience of sadness, anger, fear, and happiness lead to activation in several of the sites mentioned above, but that the pattern for each emotion is distinctive. For instance, sadness consistently activates the ventromedial prefrontal cortex, hypothalamus, and brain stem, while anger or fear activate neither the prefrontal cortex nor hypothalamus. Brain-stem activation is shared by all three emotions, but intense hypothalamic and ventromedial prefrontal activation appears specific to sadness.[15]

Third, some of these sites are also involved in the recognition of stimuli which signify certain emotions. For instance, a series of studies in my laboratory has shown that a structure known as the amygdala, which sits in the depth of each temporal lobe, is indispensable to

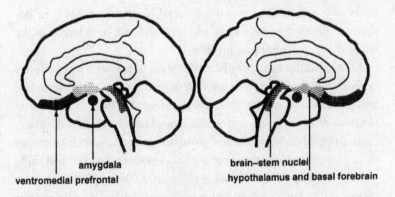

amygdala
ventromedial prefrontal

brain–stem nuclei
hypothalamus and basal forebrain

Figure 2.1. Principal emotion induction sites. Only one of these four sites is visible on the brain's surface (the ventromedial prefrontal region). The other regions are subcortical (see figure A.3 in the appendix for exact location). They are all located close to the brain's midline.

recognizing fear in facial expressions, to being conditioned to fear, and even to expressing fear. (In a parallel body of work, the studies of Joseph LeDoux and Michael Davis have shown that the amygdala is necessary for fear conditioning and revealed details of the circuitry involved in the process.[16]) The amygdala, however, has little interest in recognizing or learning about disgust or happiness. Importantly, other structures, just as specifically, are interested in those other emotions and not in fear.

The following description illustrates the fine etching of brain systems related to the production and recognition of emotion. It is but one among several examples that might be adduced to support the idea that there is no single brain center for processing emotions but rather discrete systems related to separate emotional patterns.

Have No Fear

Almost a decade ago, a young woman, to whom I shall refer as S, caught my attention because of the appearance of her brain CT scan. Unexpectedly, her scan revealed that both amygdalae, the one in the left and the one in the right temporal lobes, were almost entirely calcified. The appearance is striking. In a CT scan the normal brain shows up in myriad gray pixels, and the shade of gray defines the contours of the structures. But if a mineral like calcium has been deposited within the brain mass, the scan shows it as a bright milky white that you cannot possibly miss.

All around the two amygdalae, the brain of patient S was perfectly normal. But the amount of calcium deposition was such within the amygdalae that it was immediately apparent that little or no normal function of the neurons within the amygdalae could still take place. Each amygdala is very much a crossroads structure, with pathways from numerous cortical and subcortical regions ending in it and pathways emanating from it to just as many sites. The normal operations carried out by such profuse pathway cross signaling could simply not take place on either side of the brain of S. Nor was this a recent condition in her brain. The deposition of minerals within brain tissue takes a long time to occur and the thorough and selective job we could witness in her brain had probably taken many years to accomplish, having begun within the first years of her life. For those who are curious about the causes behind the problem, I will say that S suffers from Urbach-Wiethe disease, a rare autosomal recessive condition characterized by abnormal depositions of calcium in the skin and throat. When the brain is affected by calcium deposits, the most frequently targeted structures are the amygdalae. Those patients often have seizures, fortunately not severe, and a minor seizure was indeed the reason why S first came to our care. We were able to help her and she has not had any seizures since.

My first impression of S was of a tall, slender, and extremely pleasant young woman. I was especially curious to find out about her learning and memory ability and about her social demeanor. The reason for this curiosity was twofold. There was considerable controversy at the

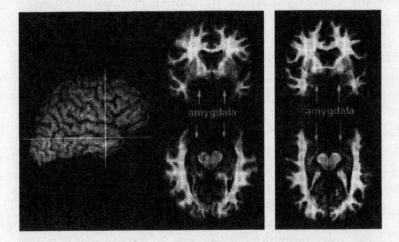

Figure 2.2. Bilateral damage of the amygdala in patient S (left panel) and normal amygdala (right panel). The sections were obtained along the two perpendicular planes shown by the white lines drawn over the brain's external surface. The black areas identified by the arrows are the damaged amygdalae. Compare with the normal amygdalae of a control brain shown in the exact same sections in the two panels on the right.

time regarding the contribution of the amygdalae to the learning of new facts, some investigators believing the amygdala was a vital partner to the hippocampus in the acquisition of new factual memory, other investigators believing it had little to contribute on that score. The curiosity regarding her demeanor was based on the fact that from studies involving nonhuman primates, it was known that the amygdala plays a role in social behaviors.[17]

I can make a long story short by telling you that there was nothing wrong whatsoever with S's ability to learn new facts. This was evident when I met her for only the second time and she clearly recognized me, smiled, and greeted me by name. Her one-shot learning of who I was, what my face looked like, and of my name was flawless. Numerous psychological tests would bear out this first impression, and that is precisely how things remain today. Years later, we were to show that a particular aspect of her learning was defective, but this had nothing to do with learning facts: it had to do with conditioning to unpleasant stimuli.[18]

Her social history, on the other hand, was exceptional. To put it in the simplest possible terms, I would say that S approached people and situations with a predominantly positive attitude. Others would actually say that her approach was excessively and inappropriately forthcoming. S was not only pleasant and cheerful, she seemed eager to interact with most anyone who would engage her in conversation, and several members of the clinical and research teams felt that the reserve and reticence one would have expected from her was simply lacking. For instance, shortly after an introduction, S would not shy away from hugging and touching. Make no mistake, her behavior caused no discomfort to anyone, but it was invariably perceived as a far cry from the standard behavior of a patient in her circumstances.

We were to learn that this very same attitude pervaded all areas of her life. She made friends easily, formed romantic attachments without difficulty, and had often been taken advantage of by those she trusted. On the other hand, she was and is a conscientious mother, and she tries hard to abide by social rules and be appreciated for her efforts. Human nature is indeed hard to describe and full of contradictions in the best of circumstances and the prime of health. It is almost impossible to do justice to it when we enter the realm of disease.

The first years of research on S yielded two important results. On the one hand, S did not have any problem learning facts. In fact, it was possible to say that her sensory perceptions, her movements, her language, and her basic intelligence were no different from those of an entirely healthy average individual in terms of elementary competence. On the other hand, her social behavior demonstrated a consistent skewing of her prevailing emotional tone. It was as if negative emotions such as fear and anger had been removed from her affective vocabulary, allowing the positive emotions to dominate her life, at least by greater frequency of occurrence if not by greater intensity. This was of special interest to me because I had noticed a similar pattern in patients with bilateral damage to the anterior sector of the temporal lobe, who, as a part of their large lesions, also had damage to the amygdalae. It was reasonable to hypothesize that their affective lopsidedness was traceable to damage in the amygdala.

All of these suppositions were to be turned into hard fact when Ralph Adolphs joined my laboratory. Using a variety of clever techniques in the investigation of several patients, some with damage to the amygdala and some with damage to other structures, Adolphs was able to determine that the affective lopsidedness was mostly caused by the impairment of one emotion: fear.[19]

Using a multidimensional scaling technique, Adolphs showed that S cannot consistently tell the expression of fear in another person's face, especially when the expression is ambiguous or other emotions are being expressed simultaneously. She has no such problem with the recognition of other facial expressions of emotion, namely, that of surprise which is, in many respects, similar in general configuration. Curiously, S, who has a remarkable gift for drawing and has good drafting skills, cannot draw a face that represents fear although she can draw faces that represent other emotions. When asked to mimic facial expressions of emotions she does so easily for the primary emotions but not for fear. Her attempts produce little change in her facial expression after which she confesses her complete failure. Again, she has no difficulty producing a facial expression of surprise. Lastly, S does not experience fear in the same way you or I would in a situation that would normally induce it. At a purely intellectual level she knows what fear is supposed to be, what should cause it, and even what one may do in situations of fear, but little or none of that intellectual baggage, so to speak, is of any use to her in the real world. The fearlessness of her nature, which is the result of the bilateral damage to her amygdalae, has prevented her from learning, throughout her young life, the significance of the unpleasant situations that all of us have lived through. As a result she has not learned the telltale signs that announce possible danger and possible unpleasantness, especially as they show up in the face of another person or in a situation. Nowhere has this been proved more clearly than in a recent study requiring a judgment of trustworthiness and approachability based on human faces.[20]

The experiment called for the judgment of one hundred human faces that had been previously rated by normal individuals as indicating varied degrees of trustworthiness and approachability. There were fifty faces that had been consistently judged as inspiring trust and fifty that were not. The selection of these faces was made by normal individuals who were asked a simple question: How would you rate this face on a scale of one to five, relative to the trustworthiness and approachability that the owner of the face inspires? Or, in other words, how eager would you be to approach the person with this particular face if you needed help?

Once the one hundred faces were properly distributed based on the ratings of the forty-six normal individuals, we turned to patients with brain damage. S was one of three patients with bilateral damage to the amygdala included in the study, but we also investigated the performance of seven patients with damage to either the left amygdala or right amygdala, three patients with damage to the hippocampus and an inability to learn new facts, and ten patients with damage elsewhere in the brain, i.e., outside the amygdala and outside the hippocampus. The results were far more remarkable than we expected.

S, along with other patients who also have damage to the amygdalae on both sides of the brain, looked at faces that you or I would consider trustworthy and classified them, quite correctly, as you or I would, as faces that one might approach in case of need. But when they looked at faces of which you or I would be suspicious, faces of persons that we would try to avoid, they judged them as equally trustworthy. The patients with damage to only one amygdala, the amnesic patients, and the other brain-damaged patients performed as normals do.

The inability to make sound social judgments, based on previous experience, of situations that are or are not conducive to one's welfare has important consequences for those who are so affected. Immersed in a secure Pollyanna world, these individuals cannot protect themselves against simple and not-so-simple social risks and are thus more vulnerable and less independent than we are. Their life histories testify to this chronic impairment as much as they testify to the paramount importance of emotion in the governance not just of simple creatures but of humans as well.

How It All Works

In a typical emotion, then, certain regions of the brain, which are part of a largely preset neural system related to emotions, send commands to other regions of the brain and to most everywhere in the body proper. The commands are sent via two routes. One route is the bloodstream, where the commands are sent in the form of chemical molecules that act on receptors in the cells which constitute body tissues. The other route consists of neuron pathways and the commands along this route take the form of electrochemical signals which act on other neurons or on muscular fibers or on organs (such as the adrenal gland) which in turn can release chemicals of their own into the bloodstream.

The result of these coordinated chemical and neural commands is a global change in the state of the organism. The organs which receive the commands change as a result of the command, and the muscles, whether the smooth muscles in a blood vessel or the striated muscles in the face, move as they are told to do. But the brain itself is changed just as remarkably. The release of substances such as monoamines and peptides from regions of nuclei in the brain stem and basal forebrain alters the mode of processing of numerous other brain circuits, triggers certain specific behaviors (for example, bonding, playing, crying), and modifies the signaling of body states to the brain. In other words, both the brain and the body proper are largely and profoundly affected by the set of commands although the origin of those commands is circumscribed to a relatively small brain area which responds to a particular content of the mental process. Now consider this: Beyond emotion, specifically described as the collection of responses I just outlined, two additional steps must take place before an emotion is *known*. The first is feeling, the imaging of the changes we just discussed. The second is the application of core consciousness to the entire set of phenomena. Knowing an emotion—feeling a feeling—only occurs at that point.

These events can be summarized by walking through the three key steps of the process:

1. Engagement of the organism by an inducer of emotion, for instance, a particular object processed visually, resulting in visual representations of the object. Imagine running into Aunt Maggie, whom you love and have not seen in a long time. Chances are you will immediately recognize Aunt Maggie, but even if you do not, or even before you do, the basic process of emotion will continue on to the next step.

2. Signals consequent to the processing of the object's image activate all the neural sites that are prepared to respond to the particular class of inducer to which the object belongs. The sites I am talking about—for instance, in the ventromedial prefrontal cortices, amygdala, and brain stem—have been preset innately, but past experience with things Maggie has modulated the manner in which they are likely to respond, for instance, the ease with which they will respond. By the way, Aunt Maggie is not traveling all over your brain in the form of a passport

photo. She exists as a visual image, arising out of neural patterns generated by the interaction of several areas in early visual cortices, largely in occipital lobes. Signals consequent to the presence of her image travel elsewhere and do their job when parts of the brain that are interested in things Maggie respond to such signals.

3. As a result of step 2, emotion induction sites trigger a number of signals toward other brain sites (for instance, monoamine nuclei, somatosensory cortices, cingulate cortices) and toward the body (for instance, viscera, endocrine glands), as previously discussed. Under some circumstances the balance of responses may favor intrabrain circuitry and engage the body minimally. This is what I have called "as if body loop" responses.

The combined result of steps 1, 2, and 3 is a momentary and appropriate collection of responses to the circumstances causing the whole commotion: for instance, Aunt Maggie in sight; or the death of a friend announced; or nothing that you can tell consciously; or, if you are a baby bird in a high nest, the image of a large object flying overhead. Take the latter example. The baby bird has no idea that this is a predatory eagle, and no conscious sense of the danger of the situation. No thought process, in the proper meaning of the term, tells the baby bird to do what it does next, which is to crouch as low as possible in the nest, as quietly as possible, such that it may become invisible to the eagle. And yet, the steps of the process that I have just described were engaged: visual images were formed in the baby bird's visual brain, some sectors of the brain responded to the *kind* of visual image the brain formed, and all the appropriate responses, chemical and neural, autonomic and motor, were engaged at full tilt. The quiet and slow tinkering of evolution has done all the thinking for the baby bird, and its genetic system has dutifully transmitted it. With a little bit of help from mother bird and earlier circumstances, the miniconcert of fear is ready to be played whenever the situation demands it. The fear response that you can see in a dog or a cat is executed in exactly the same manner, and so is the fear response you can examine in yourself when you walk at night on a dark street. That we, and at least the dog and the cat, can also come to know about the feelings caused by those emotions, thanks to consciousness, is another story.

In fact, you can find the basic configurations of emotions in simple organisms, even in unicellular organisms, and you will find yourself attributing emotions such as happiness or fear or anger to very simple creatures who, in all likelihood, have no feeling of such emotions in the sense that you or I do, creatures which are too simple to have a brain, or, having one, too rudimentary to have a mind. You make those attributions purely on the basis of the movements of the organism, the speed of each act, the number of acts per unit of time, the style of the movements, and so on. You can do the same thing with a simple chip moving about on a computer screen. Some jagged fast movements will appear "angry," harmonious but explosive jumps will look "joyous," recoiling motions will look "fearful." A video that depicts several geometric shapes moving about at different rates and holding varied relationships reliably elicits attributions of emotional state from normal adults and even children. The reason why you can anthropomorphize the chip or an animal so effectively is simple: emotion, as the word indicates, is about movement, about externalized behavior, about certain orchestrations of reactions to a given cause, within a given environment.[21]

Somewhere between the chip and your pet sits one of the living creatures that has most contributed to progress in neurobiology, a marine snail known as *Aplysia californica*. Eric Kandel and his colleagues have made great inroads in the study of memory using this very simple snail which may not have much of a mind but certainly has a scientifically decipherable nervous system and many interesting behaviors. Well, *Aplysia* may not have feelings as you or I do, but it has something not unlike emotions. Touch the gill of an *Aplysia*, and you will see the gill recoil swiftly and completely, while the heart rate of *Aplysia* goes up and it releases ink into the surroundings to confuse the enemy, a bit like James Bond when he is hotly pursued by Dr. No. *Aplysia* is emoting with a miniconcert of responses that is formally no different, only simpler, from the one that you or I could display under comparable circumstances. To the degree that *Aplysia* can represent its emotive state in the nervous system, it may have the makings of a feeling. We do not know whether *Aplysia* has feelings or not, but it is extremely difficult to imagine that *Aplysia* would know of such feelings if it does have them.[22]

SHARPENING THE DEFINITION OF EMOTION: AN ASIDE

What qualifies for an emotion? Does pain? Does a startle reflex? Neither does, but if not, why not? The closeness of these related phenomena calls for sharp distinctions but the differences tend to be ignored. Startle reflexes are part of the repertoire of regulatory responses available to complex organisms and are made up of simple behaviors (e.g., limb withdrawal). They may be included among the numerous and concerted responses that constitute an emotion—endocrine responses, multiple visceral responses, multiple musculoskeletal responses, and so on. But even the simple emotive behavior of the *Aplysia* is more complicated than a simple startle response.

Pain does not qualify for emotion, either. Pain is the consequence of a state of local dysfunction in a living tissue, the consequence of a stimulus—impending or actual tissue damage—which causes the sensation of pain but also causes regulatory responses such as reflexes and may also induce emotions on its own. In other words, emotions can be caused by the same stimulus that causes pain, but they are a different result from that same cause. Subsequently, we can come to know that we have pain and that we are having an emotion associated with it, provided there is consciousness.

When you picked up that hot plate the other day and burned the skin of your fingers, you had pain and might even have suffered from having it. Here is what happened to you, in the simplest neurobiological terms:

First, the heat activated a large number of thin and unmyelinated nerve fibers, known as C-fibers, available near the burn. (These fibers, which are distributed literally everywhere in the body, are evolution-arily old and are largely dedicated to carrying signals about internal body states, including those that will end up causing pain. They are called *unmyelinated* because they lack the insulating sheath known as myelin. Lightly myelinated fibers known as A-δ fibers travel along with C-fibers and perform a similar role. Together they are called *noci-ceptive* because they respond to stimuli that are potentially or actually damaging to living tissues.)

Second, the heat destroyed several thousand skin cells, and the de-struction released a number of chemical substances in the area.

Third, several classes of white blood cell concerned with repairing tissue damage were called to the area, the call having come from some of the released chemicals (e.g., a peptide known as substance P and ions such as potassium).

Fourth, several of those chemicals activated nerve fibers on their own, joining their signaling voices to that of the heat itself.

Once the activation wave was started in the nerve fibers, it traveled to the spinal cord and a chain of signals was produced across several neurons (a neuron is a nerve cell) and several synapses (a synapse is the point where two neurons connect and transmit signals) along the appropriate pathways. The signals went all the way into the top levels of the nervous system: the brain stem, the thalamus, and even the cerebral cortex.

What happened as a result of the succession of signals? Ensembles of neurons located at several levels of the nervous system were tem-porarily activated and the activation produced a neural pattern, a sort of map of the signals related to the injury in your fingers. The central nervous system was now in possession of multiple and varied neural patterns of tissue damage selected according to the biological specifi-cations of your nervous system and of the body proper with which it connects. The conditions needed to generate a sensation of pain had been met.

The question that I am leading to arrives at this point: Would one or all of those neural patterns of injured tissue be the same thing as *knowing* that you had pain? And the answer is, not really. Knowing that you have pain requires something else that occurs *after* the neural patterns that correspond to the substrate of pain—the nociceptive signals—are displayed in the appropriate areas of the brain stem, thalamus, and cerebral cortex and generate an image of pain, a feeling of pain. But note that the "after" process to which I am referring is not beyond the brain, it is very much in the brain and, as far as I can fathom, is just as biophysical as the process that came before. Specifically, in the example above, it is a process that interrelates neural patterns of tissue damage with the neural patterns that stand for *you*, such that yet another neural pattern can arise—the neural pattern of you knowing, which is just another name for consciousness. If the latter interrelating process does not take place, you will never know that there was tissue damage in your organism—if there is no you and there is no knowing, there is no way for you to know, right?

Curiously, if there had been no you, i.e., if you were not conscious and if there had been no self and no knowing relative to hot plates and burning fingers, the wealthy machinery of your self-*less* brain would still have used the nociceptive neural patterns generated by tissue damage to produce a number of useful responses. For instance, the organism would have been able to withdraw the arm and hand from the source of heat within hundreds of milliseconds of the beginning of tissue damage, a reflex process mediated by the central nervous system. But notice that in the previous sentence I said "organism" rather than "you." Without knowing and self, it would not have been quite "you" withdrawing the arm. Under those circumstances, the reflex would belong to the organism but not necessarily to "you." Moreover, a number of emotional responses would be engaged automatically, producing changes in facial expression and posture, along with changes in heart rate and control of blood circulation—we do not learn to wince with pain, we just wince. Although all of these responses, simple and not so simple, occur reliably in comparable situations in all conscious human beings, consciousness is not needed at all for the responses to take place. For instance, many of these responses are present even in comatose patients in whom consciousness is suspended—one of the ways in which we neurologists evaluate the state of the nervous system in an unconscious patient consists of establishing whether the patient reacts with facial and limb movements to unpleasant stimuli such as rubbing the skin over the sternum.

Tissue damage causes neural patterns on the basis of which your organism is in a state of pain. If you are conscious, those same patterns can also allow *you* to know you have pain. But whether or not you are conscious, tissue damage and the ensuing sensory patterns also cause the variety of automated responses outlined above, from a simple limb withdrawal to a complicated negative emotion. In short, pain and emotion are not the same thing.

You may wonder how the above distinction can be made, and I can give you a large body of evidence in its support. I will begin with a fact that comes from direct experience, early in my training, of a patient in whom the dissociation between *pain as such* and *emotion caused by pain* was vividly patent.[23] The patient was suffering from a severe case of refractory trigeminal neuralgia, also known as tic douloureux. This is a condition involving the nerve that supplies signals for face sensation in which even innocent stimuli, such as a light touch of the skin of the face or a sudden breeze, trigger an excruciating pain. No medication would help this young man who could do little but crouch, immobilized, whenever the excruciating pain stabbed his flesh. As a last resort, the neurosurgeon Almeida Lima, who was also one of my first mentors, offered to operate on him, because producing small lesions in a specific sector of the frontal lobe had been shown to alleviate pain and was being used in last-resort situations such as this.

I will not forget seeing the patient on the day before the operation, afraid to make any movement that might trigger a new round of pain, and then seeing him two days after the operation, when we visited him on rounds; he had become an entirely different person, relaxed, happily absorbed in a game of cards with a companion in his hospital room. When Lima asked him about the pain, he looked up and said quite cheerfully that "the pains were the same," but that he felt fine now. I remember my surprise as Lima probed the man's state of mind a bit further. The operation had done little or nothing to the sensory patterns corresponding to local tissue dysfunction that were being supplied by the trigeminal system. The mental images of that tissue dysfunction were not altered and that is why the patient could report that the pains were the same. And yet the operation had been a success. It had certainly abolished the emotional reactions that the sensory patterns of tissue dysfunction had been engendering. Suffering was gone. The facial expression, the voice, and the general deportment of this man were not those one associates with pain.

This sort of dissociation between "pain sensation" and "pain affect" has been confirmed in studies of groups of patients who underwent surgical procedures for the management of pain. More recently, Pierre Rainville, who is now an investigator in my laboratory, has shown by means of a clever manipulation using hypnosis that pain sensation and pain affect are clearly separable. Hypnotic suggestions designed to influence pain affect specifically without altering pain sensation modulated cerebral activity within the cingulate cortex, the same overall region that neurosurgeons can damage to alleviate suffering from chronic and intractable pain. Rainville has also shown that when hypnotic suggestions were aimed at pain sensation rather than at the emotions associated with pain, not only were there changes in *both* unpleasantness and intensity ratings, but also there were changes in S_I (the primary somatosensory cortex) and the cingulate cortex.[24] In brief: hypnotic suggestions aimed at the emotions that follow pain rather than at pain sensation reduced emotion but not pain sensation and also caused functional changes in cingulate cortex only; hypnotic suggestions aimed at pain sensation reduced *both* pain sensation and emotion, and caused functional changes in S_I and in the cingulate cortex. Perhaps you have had the direct experience of what I am describing if you have ever taken beta-blockers to treat a heart-rhythm problem or if you have taken a tranquilizer such as Valium. Those medications reduce your emotional reactivity, and should you also have pain at the time, they will reduce the emotion caused by pain.

We can verify the different biological status of pain and emotion by considering how different interventions interfere with one but not the other. For instance, the stimuli that cause pain can be specifically reduced or blocked by analgesia. When the transmission of signals leading to the representation of tissue dysfunction is blocked, neither pain nor emotion ensue. But it is possible to block emotion and *not* pain. The would-be emotion caused by tissue damage can be reduced by appropriate drugs, e.g., Valium or beta-blockers, or even by selective surgery. The perception of tissue damage remains but the blunting of emotion removes the suffering that would have accompanied it.

And what about pleasure? Is pleasure an emotion? Again, I would prefer to say it is not, although, just like pain, pleasure is intimately related to emotion. Like pain, pleasure is a constituent quality of certain emotions as well as a trigger for certain emotions. While pain is associated with negative emotions, such as anguish, fear, sadness, and disgust, whose combination commonly constitutes what is called suffering, pleasure is associated with many shades of happiness, pride, and positive background emotions.

Pain and pleasure are part of biological design for obviously adaptive purposes, but they do their job in very different circumstances. Pain is the perception of a sensory representation of local living-tissue dysfunction. In most circumstances when there is actual or impending damage to living tissues there arise signals that are transmitted both chemically and via nerve fibers of the C and A-δ type, and appropriate representations are created in the central nervous system, at multiple levels. In other words, the organism is designed to respond to the actual or threatened loss of integrity of its tissue with a particular type of signaling. The signaling recruits a host of chemical and neural responses all the way from local reactions of white blood cells, to reflexes involving an entire limb, to a concerted emotional reaction.

Pleasure arises in a different setting. Turning to the simple example of pleasures associated with eating or drinking, we see that pleasure is commonly initiated by a detection of imbalance, for instance, low blood sugar or high osmolality. The unbalance leads to the state of hunger or thirst (this is known as a motivational and drive state), which leads in turn to certain behaviors involving the search for food or water (also part and parcel of the motivational and drive state), which leads to the eventual acts of eating or drinking. The control of these several steps involves many functional loops, at different hierarchies, and requires the coordination of internally produced chemical substances and neural activity.[25] The pleasurable state may begin during the search process, in anticipation of the actual goal of the search, and increase as the goal is achieved.

But between the cup and the lip many a slip. A search for food or drink that takes too long or is unsuccessful will not be accompanied by pleasure and positive emotions at all. Or, if in the course of a successful search, an animal is prevented from actually achieving its goal, the thwarting of the consummation may actually cause anger. Likewise, as I noted in my comment on Greek tragedy, the alleviation or suspension of a state of pain may cause the emergence of pleasure and positive emotions.

The point to retain here is the possible interrelationship between pain and pleasure and the attending emotions, as well as the fact that they are not the mirror image of each other. They are different and asymmetric physiological states, which underlie different perceptual qualities destined to help with the solution of very different problems. (The duality of pain and pleasure should not make us overlook the fact that there are more than two emotions, some of which are aligned with pain and some with pleasure, mostly the former. The apparent symmetry of this deep division vanishes as behaviors become more complex in evolution.) In the case of pain, the problem is coping with the loss of integrity of living tissue as a result of injury, be it internally caused by natural disease or externally induced by the attack of a predator or by an accident. In the case of pleasure, the problem is to lead an organism to attitudes and behaviors that are conducive to the maintenance of its homeostasis. Curiously, pain, which I regard as one of the main determinants of the course of biological and cultural evolution, may have begun as an afterthought of nature, an attempt to deal with a problem that has already arisen. I used to think of pain as putting a good lock on the door after a house has been robbed, but Pierre Rainville has suggested a better metaphor to me: putting a *body-guard* in front of the house while you repair the broken window. After all, pain does not result in preventing yet another injury, at least not immediately, but rather in protecting the injured tissue, facilitating tissue repair, and avoiding infection of the wound. Pleasure, on the other hand, is all about forethought. It is related to the clever anticipation of what can be done *not* to have a problem. At this basic level, nature found a wonderful solution: it seduces us into good behavior.

Pain and pleasure are thus part of two different genealogies of life regulation. Pain is aligned with punishment and is associated with behaviors such as withdrawal or freezing. Pleasure, on the other hand, is aligned with reward and is associated with behaviors such as seeking and approaching.

Punishment causes organisms to close themselves in, freezing and withdrawing from their surroundings. Reward causes organisms to open themselves up and out toward their environment, approaching it, searching it, and by so doing increasing both their opportunity of survival and their vulnerability.

This fundamental duality is apparent in a creature as simple and presumably as nonconscious as a sea anemone. Its organism, devoid of brain and equipped only with a simple nervous system, is little more than a gut with two openings, animated by two sets of muscles, some circular, the others lengthwise. The circumstances surrounding the sea anemone determine what its entire organism does: open up to the world like a blossoming flower—at which point water and nutrients enter its body and supply it with energy—or close itself in a contracted flat pack, small, withdrawn, and nearly imperceptible to others. The essence of joy and sadness, of approach and avoidance, of vulnerability and safety, are as apparent in this simple dichotomy of brainless behavior as they are in the mercurial emotional changes of a child at play.

THE SUBSTRATE FOR THE REPRESENTATION OF EMOTIONS AND FEELINGS

There is nothing vague, nothing elusive, nothing nonspecific, about the collection of responses I have just described as constituting an emotion. The substrate for the representation of emotions is a collection of neural dispositions in a number of brain regions located largely in subcortical nuclei of the brain stem, hypothalamus, basal forebrain, and amygdala. In keeping with their dispositional status, these representations are implicit, dormant, and not available to consciousness. They exist, rather, as potential patterns of activity arising within neuron ensembles. Once these dispositions are activated, a number of consequences ensue. On the one hand, the pattern of activation represents, within the brain, a particular emotion as neural "object." On the other, the pattern of activation generates explicit responses that modify both the state of the body proper and the state of other brain regions. By so doing, the responses create an emotional state, and at that point, an external observer can appreciate the emotional engagement of the organism being observed. As for the internal state of the organism in which the emotion is taking place, it has available both the emotion as neural object (the activation pattern at the induction sites) and the sensing of the consequences of the activation, a feeling, provided the resulting collection of neural patterns becomes images in mind.

The neural patterns which constitute the substrate of a feeling arise in two classes of biological changes: changes related to body state and changes related to cognitive state. The changes related to body state are achieved by one of two mechanisms. One involves what I call the "body loop." It uses both humoral signals (chemical messages conveyed via the bloodstream) and neural signals (electrochemical messages conveyed via nerve pathways). As a result of both types of signal, the body landscape is changed and is subsequently represented in somatosensory structures of the central nervous system, from the brain stem on up. The change in the representation of the body landscape can be partly achieved by another mechanism, which I call the "as if body loop." In this alternate mechanism, the representation of body-related changes is created directly in sensory body maps, under the control of other neural sites, for instance, the prefrontal cortices. It is "as if" the body had really been changed but it was not.

The changes related to cognitive state are no less interesting. They occur when the process of emotion leads to the secretion of certain chemical substances in nuclei of the basal forebrain, hypothalamus, and brain stem, and to the subsequent delivery of those substances to several other brain regions. When these nuclei release certain neuromodulators (such as monoamines) in the cerebral cortex, thalamus, and basal ganglia, they cause several significant alterations of brain function. The full range of alterations is not completely understood yet, but here are the most important I envision: (1) the induction of specific behaviors such as those aimed at generating bonding, nurturing, exploration, and playing; (2) a change in the ongoing processing of body states such that body signals may be filtered or allowed to pass, be selectively inhibited or enhanced, and their pleasant or unpleasant quality modified; and (3) a change in the mode of cognitive processing such that, for example, the rate of production of auditory or visual images can be changed (from slow to fast or vice versa) or the focus of images can be changed (from sharply focused to vaguely focused); changes in rate of production or focus are an integral part of emotions as disparate as those of sadness or elation.

Assuming that all the proper structures are in place, the processes reviewed above allow an organism to undergo an emotion, exhibit it, and image it, that is, feel the emotion. But nothing in the above review indicates how the organism could know that it was feeling the emotion it was undergoing. For an organism to know that it has a feeling, it is necessary to add the process of consciousness in the aftermath of the processes of emotion and feeling. In the chapters ahead I give you my idea of what consciousness is and of how it may work so that we can "feel" a feeling.

Chapter Three

Core Consciousness

STUDYING CONSCIOUSNESS

It is fine for us scientists to bemoan the fact that consciousness is an entirely personal and private affair and that it is not amenable to the third-person observations that are commonplace in physics and in other branches of the life sciences. We must face the fact, however, that this is the situation and turn the hurdle into a virtue. Above all, we must not fall in the trap of attempting to study consciousness exclusively from an external vantage point based on the fear that the internal vantage point is hopelessly flawed. The study of human consciousness requires both internal and external views.

Although the investigation of consciousness is condemned to some indirectness, this limitation is not restricted to consciousness. It applies to all other cognitive phenomena. Behavioral acts—kicks, punches, and words—are nice expressions of the private process of mind, but they are not the same thing. Likewise, electroencephalograms and functional MRI scans capture correlates of the mind but those correlates are not the mind. Inevitable indirectness, however, is not equivalent to eternal ignorance about mental structures or about the underlying neural mechanisms. The fact that mental images are accessible only to their owner organism does not preclude their characterization, does not deny their reliance on organic substance, and does not prevent our gradual closing in on the specifications of that substance. This may cause some worry to purists raised on the idea that what another person cannot see is not to be trusted scientifically, but it really should not worry anyone. This state of affairs should not prevent us from treating subjective phenomena scientifically. Whether one likes it or not, *all* the contents in our minds are subjective and the power of science comes from its ability to verify objectively the consistency of many individual subjectivities.

Consciousness happens in the interior of an organism rather than in public, but it is associated with a number of public manifestations. Those manifestations do not describe the internal process in the same direct way that a spoken sentence translates a thought, yet there they are, available to observation, as correlates and telltale signs of the presence of consciousness. Based on what we know about private human minds and on what we know and can observe of human behavior, it is possible to establish a three-way link among: (1) certain external manifestations, e.g., wakefulness, background emotions, attention, specific behaviors; (2) the corresponding internal manifestations of the human being having those behaviors as reported by that human being; and (3) the internal manifestations that we, as observers, can verify in ourselves when we are in circumstances equivalent to those of the observed individual. This three-way linkage authorizes us to make reasonable inferences about human private states based on external behavior.[1]

The solution of the method problem posed by the privacy of consciousness relies on a natural human ability, that of theorizing constantly about the state of mind of others from observations of behaviors, reports of mental states, and counterchecking of their correspondences, given one's own comparable experiences. As a student of mind and behavior I turned a pastime—curiosity about the minds of others—into a professional activity, which simply means that I was obsessive about it and took notes.

Curiously, compared with the specialists, the popular culture seems to have fewer problems with the private perspective of consciousness, as shown brilliantly in Woody Allen's *Deconstructing Harry*. Perhaps you have seen the film, but if not, here is my report on what happens. In the middle of a movie-within-a-movie scene, which describes the shooting of a film scene, the cameraman realizes that the image of the actor he is filming is fuzzy. Naturally, he first attributes the problem to his own error in controlling the focus, and after he fails to correct it, he begins worrying that the focusing mechanism may be out of order. But the mechanism is fine, and since there is no improvement, the cameraman now worries about the state of the lens. Could it be dirty and so cause the fuzziness? Yet, the lens turns out to be fine, too, and perfectly clean. In the midst of the ensuing commotion, everyone suddenly realizes that the problem does not have anything to do with the camera at all but with the actor in question (Mel, played by Robin Williams). It is the actor himself who is out of focus! He is *intrinsically* fuzzy, and everyone looking at him sees a blurred image; everyone looking at anything else but Mel sees a clear image. The actor of this movie-within-a-movie has been struck by a disease that makes all those around him, including his perplexed family and his physician, see him out of focus.

The reason why the audience laughs has to do with the patent absurdity of the idea, with the violation of a property fundamental to consciousness: its personal, private, first-person view of things. Fuzziness and out-of-focusness are not properties of objects—except in a metaphorical sense. Even when a screen is interposed between you and an object and modifies its perception, i.e., when the lenses of your glasses are dirty, the fuzziness is not *in* the object. Fuzziness and out-of-focusness are very much a part of our conscious perspective in perception. In normal circumstances, fuzziness and out-of-focusness occur *within* a person's organism, due to a number of possible causes arising at a variety of physiological levels, all the way from the eye to the pathways that transmit signals to the brain, to the brain itself. Other persons in the vicinity of he-who-seems-fuzzy-to-me do not share my fuzziness and my out-of-focusness. The scene succeeds because no one can bring Mel into focus. Fuzziness has become an external property of a living being rather than the personally constructed feature of an observation.

The contemporary approach to studying the biological basis of the private human mind involves two steps. The first step consists of observing and measuring the actions of an experimental subject, or collecting and measuring the reports of internal experience offered by a subject, or both. The second step consists of relating the collected evidence to the measured manifestation of one of the neurobiological phenomena we are beginning to understand, at the level of molecules, neurons, neural circuits, or systems of circuits. The approach is based on the following assumptions: that the processes of the mind, including those of consciousness, are based on brain activity; that the brain is a part of a whole organism with which it interacts continuously; and that we, as human beings, in spite of remarkable individual traits that make each of us unique, share similar biological characteristics in terms of the structure, organization, and function of our organisms.

The limits of the first-pass solution outlined above can widen remarkably when we transfer the approach to human beings with neurological disease who develop impairments of mind and behavior caused by brain damage and selective brain dysfunction—the sort of problem that arises, for instance, as a result of a stroke. This approach, which is known as the lesion method, allows us to do for consciousness what we have long been doing for vision, language, or memory: investigate a breakdown of behavior, connect it to the breakdown of mental states (cognition), and connect both to a focal brain lesion (an area of circumscribed brain damage) or to an abnormal record of electrical activity assessed with an electroencephalogram or evoked electrical potentials (a brain-wave test) or an abnormality in a functional-imaging scan (such as PET or fMR). A population of neurological patients gives us opportunities that observations in normals alone do not. It gives us probes in terms of disordered behavior and mind as well as probes in terms of anatomically identifiable sites of brain dysfunction with which we can study many aspects of mind, especially those aspects that are less transparent. Armed with the ensuing evidence, it is possible to submit hypotheses to test, support them or modify them according to the results, and test perfected hypotheses in yet other neurological patients or healthy controls.

The investigation of patients with neurological disease has shaped my views on consciousness more than any other source of evidence. Before I reflect on my observations of neurologic patients with impaired consciousness, however, a word about the telltale external manifestations of consciousness is in order.

THE MUSIC OF BEHAVIOR AND THE EXTERNAL MANIFESTATIONS OF CONSCIOUSNESS

The consistent and predictable external manifestations of consciousness are readily identifiable and measurable. For instance, we know that organisms in a normal state of consciousness are awake, are attentive to stimuli in their surroundings, and behave in a manner adequate to the context and to what we imagine their purpose to be. Adequate behavior includes both the background emotions I described earlier as well as specific actions or specific emotions related to the specific events or stimuli occurring in a given scene. An expert observer can assess these correlates of consciousness over a relatively short period of time (perhaps as short as ten minutes if the circumstances are propitious, although I must add that experts can be fooled). The presence or absence of wakefulness can be established by direct observation of the organism—the eyes must be open, the muscles must have tone enough to permit movement. The ability to attend to stimuli can be established from the organism's ability to orient to stimuli, and we can observe eye movements, head movements, and patterns of limb and whole-body movement as the organism responds to varied sensory stimuli and interacts in an environment. The presence of background emotion can be established from the nature of facial expressions and from the dynamic profile of limb movements and posture. The purposefulness and adequacy of behavior can be assessed by taking into account the context of the situation, whether natural or experimental, and determining whether the organism's responses to stimuli and the organism's self-initiated actions are appropriate to that context.

Although all of these manifestations can be elicited by appropriate stimuli, observed, videotaped, and measured with various devices, I must emphasize that the qualitative judgments of the trained observer are an essential tool in the analysis of behavior. What confronts the observer is decomposable by expert analysis but is first and foremost a composite, a concurrence of contributions in time, played out in a single organism and connected, in some fashion, by a single goal.

It may be helpful to think of the behavior of an organism as the performance of an orchestral piece whose score is being invented as it goes along. Just as the music you hear is the result of many groups of instruments playing together in time, the behavior of an organism is the result of several biological systems performing concurrently. The different groups of instruments produce different kinds of sound and execute different melodies. They may play continuously throughout a piece or be absent at times, sometimes for a number of measures. Likewise for the behavior of an organism. Some biological systems produce behaviors that are present continuously, while others produce behaviors that may or may not be present at a given time. The principal ideas I wish to highlight here are: First, that the behavior we observe in a living organism is not the result of one simple melodic line but rather the result of a concurrence of melodic lines at each time unit you select for the observation; if you were a conductor looking at the imaginary musical score of the organism's behavior, you would see the different musical parts joined vertically at each measure. Second, that some components of behavior are always present, forming the continuous base of the performance while others are present only during certain periods of the performance; the "behavioral score" would note the entrance of a certain behavior at a certain measure and the end of it some measures later, just as the conductor's score notes the beginnings and ends of the solo piano parts within the movements of a concerto. Third, that in spite of the various components, the behavioral product of each moment is an integrated whole, a fusion of contributions not unlike the polyphonic fusion of an orchestral performance. Out of the critical feature I am describing here, concurrence in time, something emerges that is not specified in any of the parts.

As we consider human behavior in the pages ahead, I ask you to think of several parallel lines of performance unfolding in time. Wakefulness, background emotion, and low-level attention will be there continuously; they are present from the moment of awakening to the moment when you fall asleep. Specific emotions, focused attention, and particular sequences of actions (behaviors) will appear from time to time, as appropriate for the circumstances. Likewise for verbal reports, which are a variety of behavior.

Now, consider an extension of this metaphor into the mind of the person whose performance we are observing. I propose that there is also an orchestral score in the private mind, only now the concurrent stacking of musical parts corresponds to mental streams of images. Those streams are largely the internal and cognitive counterpart of the behaviors we observe. Some images occur a fraction of time earlier than those behaviors do, e.g., the mental image of an idea we are about to express in a sentence. Other images occur immediately after, e.g., the feeling of the emotion we just exhibited. There are, of course, musical parts for the state of being awake and continuously making images as well as for the representation of specific objects, events, and words denoting them; there is also a part for the feelings of the varied emotions the organism is exhibiting. There is, however, one other part in the internal orchestral score for which there is no precise external counterpart: that part is the sense of self, the critical component of any notion of consciousness.

In the context of this metaphor, we can imagine the sense of self as an additional part which informs the mind, nonverbally, of the very existence of the individual organism in which that mind is unfolding and of the fact that the organism is engaged in interacting with particular objects within itself or in its surroundings. This knowledge alters the course of the mental process and the course of external behavior. Its private presence, which is directly available only to its owner, can be inferred by an external observer from the influence it exerts on external behaviors, rather than from its own flagship behavior. Wakefulness, background emotion, and low-level attention are thus external signs of internal conditions that are compatible with the occurrence of consciousness. On the other hand, specific emotions, sustained and focused attention, and targeted behaviors appropriate to the context over extended periods of time are a good indication that consciousness is indeed occurring in the subject we observe, even if we, as external observers, cannot observe consciousness directly.

Table 3.1. The Behavioral Score

verbal report
specific actions
specific emotions
focused attention
low-level attention
background emotions
wakefulness

Wakefulness

Wakefulness and consciousness tend to go together, although the coupling can be broken in two exceptional circumstances. One exception occurs when we are in the state of dream sleep. We are obviously not awake during dream sleep and yet we have some consciousness of the events taking place in the mind. The memory we form of the last dream fragments before we wake up indicates that some consciousness was "on." Another dramatic reversal of the usual coupling can also occur: we can be awake and yet be deprived of consciousness. Fortunately, the latter only happens in the neurological conditions I am about to discuss.

Wakefulness is best described from watching the transition from sleep to wakefulness. The indelible picture of that transition that always comes to my mind is that of Winnie in Beckett's *Happy Days* when the bell rings at the beginning of the first act: Winnie opens her eyes to the audience and declares, "Another heavenly day." On she goes, like a morning sunrise, in a state which will permit her brain to form images of her surroundings: her bag, her toothbrush, the rustling sounds of Willie, her body, which, she tells us, does not have much pain that day, "hardly any." Wakefulness stops at the end of Winnie's day when the bell rings to close the first act.

When wakefulness is removed, dream sleep aside, consciousness is removed. Examples of this pairing are dreamless sleep, anesthesia, and coma. But wakefulness is not the same as consciousness. In the wakeful state the brain and mind are "on," and images of the organism's interior as well as the organism's environment are being formed. Reflexes can be engaged, of course (neither consciousness nor wakefulness is needed for reflex activity), and low-level attention can be driven to stimuli that conform to the basic needs of the organism. And yet, consciousness may be absent. Patients with some neurological conditions discussed in this chapter are awake and yet lack what core consciousness would have added to their thought process: images of knowing centered on a self.

Attention and Purposeful Behavior

There is more to Winnie's behavior than just wakefulness. She orients herself toward objects and concentrates on them as needed. Eyes, head, neck, torso, and arms move about in a coordinated dance which establishes an unequivocal relationship between Winnie and certain stimuli in her surroundings: the bag, the toothbrush, Willie's rustling behind her. Presence of attention toward an external object usually signifies the presence of consciousness, though not necessarily. Patients in so-called akinetic mutism, who have abnormal consciousness, can pay *fleeting* and low-level attention to a *salient* event or object, for instance, an observer calling their name. Attention only betrays the presence of normal consciousness when it can be *sustained* over a substantial period of time relative to the objects that are necessary for appropriate behavior in a given context—this means many minutes and hours rather than seconds. In other words, extended time and a focusing on appropriate objects define the sort of attention that is indicative of consciousness.

Lack of manifest attention toward an external object does not necessarily deny the presence of consciousness and may instead indicate that attention is directed toward an internal object. Absentminded professors and daydreaming adolescents exhibit this "symptom" all the time. Fortunately, the condition is most transient. Complete and sustained failure of attention is associated with the dissolution of consciousness, as happens in drowsiness, confusional states, or stupor.

Conscious creatures concentrate on certain objects and are attentive to certain stimuli, something that matches quite well our own view from within when we think about what goes on in our mind in comparable situations. We can all agree that attention and consciousness are related, but the nature of the relationship is a matter for debate. My view is that both consciousness and attention occur in levels and grades, they are not monoliths, and they influence each other in a sort of upward spiral. Low-level attention precedes core consciousness; it is needed to engage the processes that generate core consciousness. But the process of core consciousness results in driving higher-level attention toward a focus. When I attend to an acquaintance who has just turned up in my office, I do so under the influence of core consciousness. I could only have generated that consciousness because my organism was directed by low-level automated attention to process certain features of the environment that are important for organisms like mine, namely, moving creatures with human faces. As the processing continued, core consciousness helped focus attention on the particular object that engaged the organism in the first place.

But back to Winnie. Next you notice that she behaves purposefully toward the stimuli on which she concentrates. She might not—Winnie being a character in a Beckett play—but she does. In effect, her behavior is part of an immediately recognizable plan that could only have been formulated by an organism knowledgeable about its past, present, and anticipated future. The behavior is consonant with such a plan over a long period of time—hours, in fact. The sustained purposefulness and adequateness of her behavior require the presence of consciousness even if consciousness does not guarantee purposeful and adequate behavior: perfectly conscious idiots may behave quite inadequately.

Something especially noteworthy about such sustained and adequate behaving is that specific behaviors are accompanied by a flow of emotional states as part of their unfolding. The background emotions that we discussed in the previous chapter continuously underscore the subject's actions. Telltale signals include the overall body posture and the range of motion of the limbs relative to the trunk; the spatial profile of limb movements, which can be smooth or jerky; the speed of motions; the congruence of movements occurring in different body tiers such as face, hands, and legs; and last and perhaps most important, the animation of the face. Even when the observed subject speaks, emotional aspects of the communication are separate from the content of the words and sentences spoken. Words and sentences, from the simple "Yes," "No," and "Hello" to "Good Morning" or "Good-bye," are usually uttered with a background emotional inflection. The inflection is an instance of prosody, the musical, tonal accompaniment to the speech sounds that constitute the words. Prosody can express not just background emotions, but specific emotions as well. For instance, you can tell someone, in the most loving tone, "Oh! Go away!" and you can also say, "How nice to see you" with a prosody that unmistakably registers indifference.

Moreover, specific emotions often succeed stimuli or actions that seemingly motivate them in the subject, as judged from the perspective of the observer. In effect, normal human behavior exhibits a continuity of emotions induced by a continuity of thoughts. The contents of those thoughts, and there are usually parallel and simultaneous contents, include objects with which the organism is actually engaged or objects recalled from memory as well as feelings of the emotions that have just occurred. In turn, many of these "streams" of thought—of actual objects, of recalled objects, and of feelings—can induce emotions, from background to secondary, with or without our cognizance. The continuous exhibition of emotion derives from this overabundance of inducers, known and not known, simple and not so simple.

The continuity of the melodic line of background emotion is an important fact to consider in our observation of normal human behavior. When we observe someone with intact core consciousness, well before any words are spoken, we find ourselves presuming the subject's state of mind. Whether correct or not, some of the presumptions are based on a continuity of emotional signals available in the subject's behavior.

A note of caution on confusing terminology: On occasion, terms such as *alertness* and *arousal* are used as synonyms of *wakefulness, attention,* and even of *consciousness,* but they should not be. *Alertness* is often used instead of *wakefulness,* as when you say that you feel "quite alert" or that you think somebody is. For my purposes, the term *alertness* should signify that the subject is not just awake but apparently disposed to perceive and act. The proper meaning of *alert* is somewhere between "awake" and "attentive."

The term *arousal* is easier to define. It denotes the presence of signs of autonomic nervous system activation such as changes in skin color (rubor or pallor), behavior of skin hair (hair standing on end!), diameter of the pupils (larger or smaller), sweating, sexual erection, and so on, which are reasonably covered by lay terms such as *excitement*. One can be awake, alert, and fully conscious without being "aroused" in this sense, but we all know that our organisms can be "aroused" in this sense during sleep, when we are not awake, attentive, or conscious. Even comatose patients can be aroused, only they do not know it. Tricky, isn't it?

STUDYING CONSCIOUSNESS FROM ITS ABSENCE

You may wonder how we can comment, from a personal perspective, on the absence of consciousness, considering that the absence of knowing and self should preclude our experience of that absence. The answer is that we come close to experiencing the absence of consciousness in a few circumstances. Consider the brief moments during which we come to awareness after an episode of loss of consciousness caused by fainting or anesthesia; or, in a more benign sort of way, the fleeting moments which precede fully waking up from the deep compensatory sleep that follows fatigue. In those transitional instants we have a glimpse of the impoverished mental state that preceded them. Images are being formed of people and objects and places around us, and yet, for a brief period which may seem all too long, the sense of self is missing and no individual ownership of thought is apparent. A split second later our sense of self is "on," and yes, we vaguely surmise that the images belong to us but not all the details fit clearly yet. It takes a while longer for the autobiographical self to be reinstated as a process and for the situation to be perfectly explained.

The question remains, however, as to how we can possibly glimpse such a state of nonconscious mental impoverishment when we were not really conscious during that state. We certainly have such glimpses, and I suspect that the reason why we do is that we lack, in those transitional instants, the memory of any experience of the instants that came immediately before the transition. Our conscious experience normally includes a brief memory of what we sense as "the just before," which is attached to what we innocently think is the "now." That memory describes the sense of a self to whom some knowledge is being attributed. Immediately upon awakening, however, the brief memory that would have preserved the previous instant for the benefit of the current instant is not available, for the good reason that there was no conscious experience to be memorized. Our introspection of these anomalous states, then, reveals an important fact: the continuity of normal consciousness requires a brief memory, in the order of a fraction of a second, a trivial achievement for the human brain whose regular short-term memory for facts can last about sixty seconds.

The most extreme varieties of impaired consciousness—coma, persistent vegetative state, deep sleep, deep anesthesia—afford little opportunity for behavior analyses because nearly all manifestations in the "behavioral score" we discussed are abolished.[2] Correspondingly, nearly all the internal manifestations in the "cognitive score" are presumed to be abolished as well. The notion that consciousness phenomena and even mind phenomena are suspended in such situations is an intuition based on solid reflections on our own condition and on equally solid observations of the behavior of others. The notion is also fully supported by the rare but extremely valuable reports of persons who return to consciousness after being in coma. They can recall the descent into the nothingness of coma—much as we can recall the induction of general anesthesia—and the return to knowingness, but nothing at all is recalled of the intervening period, which can span weeks or months. It is legitimate to assume, given all the evidence, that little or nothing was in fact going on in the mind in such circumstances.[3]

Two other groups of patients, however, afford extensive opportunity for behavioral analyses and stand out in terms of the influence that their study had on my thinking about consciousness. One group is made up of patients with a complicated phenomenon known as *epileptic automatism*. The other group brings together patients who, as a result of a variety of neurologic diseases, develop a condition known by the blanket term *akinetic mutism*. In both groups, core consciousness and extended consciousness are profoundly affected, and yet not all of the behaviors described in the "behavioral score" are abolished, thus leaving room for some intervention by the observer and for the analysis of a residual performance.[4]

EPILEPTIC AUTOMATISMS CAN be like a scalpel and separate consciousness from the things that are in consciousness. Automatisms can appear as part of seizures or immediately following seizures. The episodes that interest me the most are associated with absence seizures, although automatisms are also seen in association with so-called temporal-lobe seizures. Absence seizures are one of the main varieties of epilepsy, in which consciousness is momentarily suspended along with emotion, attention, and adequate behavior. The disturbance is accompanied by a characteristic electrical abnormality in the EEG. Absence seizures are of great value to the student of consciousness, and the typical variety of absence seizure is in fact one of the most pure examples of loss of consciousness—the term *absence* is shorthand for "absence of consciousness." The absence automatism that follows an especially long absence seizure is perhaps the purest example of all.

If you were talking to someone prone to absence seizures and absence automatisms, here is what might happen if an episode were to begin. Suddenly, while having a perfectly sensible conversation, the patient would interrupt himself in midsentence, freeze whatever other movement he was performing, and stare blankly, his eyes focused on nothing, his face devoid of any expression—a meaningless mask. The patient would remain awake. The muscular tone would be preserved. The patient would not fall, or have convulsions, or drop whatever he was holding in his hand. This state of suspended animation might last for as little as three seconds—a far longer time than you imagine when you are watching it—and for as long as tens of seconds. The longer it lasts, the more likely it is that absence proper will be followed by absence automatism, which, once again, can take a few seconds or many. As the automatism starts, the events become even more intriguing. The situation is not unlike the unfreezing of film images when you release a freeze-frame control or when the jammed projector in a movie house gets to be unjammed. The show goes on. As the patient unfreezes he looks about, perhaps not at you but at something nearby, his face remains a blank, with no sign of a decipherable expression, he drinks from the glass on the table, smacks his lips, fumbles with his clothes, gets up, turns around, moves toward the door, opens it, hesitates just outside the threshold, then walks down the hallway. By this time you would have got up and followed him so that you might witness the end of the episode. One of several scenarios might unfold. In the most likely scenario, the patient might stop and stand somewhere in the hallway, appearing confused; or he might sit on a bench, if there were one. But the patient might possibly enter another room or continue walking. In the most extreme variety of such episodes, in what is known as an "epileptic fugue," the patient might even get out of the building and walk about in a street. To a good observer he would have looked strange and confused, but he might get by without any harm coming to him. Along the trajectory of any of these scenarios, most frequently within seconds, more rarely within a few minutes, the automatism episode would come to an end and the patient would look bewildered, wherever he would be at that moment. Consciousness would have returned as suddenly as it had disappeared, and you would have to be there to explain the situation to him and bring him back to where the two of you were before the episode began.

The patient would have no recollection whatsoever of the intervening time. The patient would not know then and not know ever what his organism had been doing during the episode. After an episode ends, such patients have no recollection of what went on during the seizure or during the extension of the seizure in the automatism period. They do remember what went on before the seizure and can retrieve those contents from memory, a clear indication that their learning mechanisms were intact prior to the seizure. They immediately learn what goes on after the seizure ends, a sign that the seizure did not produce a permanent impairment of learning. But the events that occurred during the period of seizure have not been committed to memory or are not retrievable if they have.

Were you to have interrupted the patient at any point during the episode, he would have looked at you in utter bewilderment or perhaps with indifference. He would not have known who you were, spontaneously or upon specific questioning; he would not know who he was or what he was doing; and he might have simply kept you away with a vague gesture, hardly looking at you. The contents that make up a conscious mind would have been missing, and this could no more lead to a verbal report than to a highly intelligent action. He would have remained awake and attentive enough to process the object that came next into his perceptual purview, but inasmuch as we can deduce from the situation, that is all that would go on in the mind. There would have been no plan, no forethought, no sense of an individual organism wishing, wanting, considering, believing. There would have been no sense of self, no identifiable person with a past and an anticipated future—specifically, no core self and no autobiographical self.

In such circumstances, the presence of an object promotes the next action and that action may be adequate within the microcontext of the moment—drinking from a glass, opening a door. But that action, and other actions, will not be adequate in the broader context of circumstances in which the patient is operating. As one watches actions unfold, one realizes that they are devoid of ultimate purpose and are inappropriate for an individual in that situation.

There would have been, however, unmistakable wakefulness: the eyes would have been open; muscle tone, maintained. There would have been some ability to create neural patterns and presumably images: the objects around the patient had to be sufficiently mapped in visual or tactile terms so that he could execute actions successfully. And there would also have been attention, not high-level attention like we are having at this moment, but attention enough so that the perceptual and motor devices of the organism could turn to a particular object long enough and well enough for sensory images to be properly formed and movements to be executed with accuracy relative to those images, e.g., the visual image of a wall; the tactile image of the glass from which the patient could drink.

In other words, the patient would have had some elementary aspects of mind, would have had some contents in that mind pertaining to the objects surrounding him, but he would not have had a normal consciousness. He would not have developed, in parallel with the image of the objects surrounding him, an image of knowing centered on a self; an enhanced image of the objects he was interacting with; a sense of the appropriate connection to what went on before each given instant or what might happen in the instant ahead.

The dissociation between impaired consciousness and the ability to form neural patterns for objects, surprising as it may seem, is also borne out by intriguing new evidence. A patient in persistent vegetative state, a lighter form of coma in which there are signs of wakefulness but consciousness is gravely impaired, was studied with a functional imaging scan during which photographs of familiar human faces were projected onto her retinas. The result was activation of a region in the occipitotemporal cortices known to be activated by the perception of faces in normal, awake, and conscious persons. Thus even without consciousness, the brain can process sensory signals across varied neural stations and cause activation of at least some of the areas usually involved in the processes of perception.[5]

Observing an episode of absence automatism you would have watched the elaborate behaviors of an organism deprived of all extended consciousness and of everything but perhaps the dimmest form of core consciousness. One can only try to imagine the remains of a mind from which self and knowing have been removed, perhaps a mind strewn with images of things to be known but never really known, with things not really owned—stripped of the engine for deliberate action.

Let me conclude by commenting on the fact that emotion was missing throughout the episode. The suspension of emotion is an important sign in absence seizures and in absence automatisms. Emotion is also missing in the akinetic mutisms described in the next section. The lack of emotion—no background emotions and no specific emotions—is conspicuous, but it has not been highlighted in the relevant literature. As I reflect on this finding, many years after I first noted it, I venture that absence of emotion is a reliable correlate of defective core consciousness, perhaps as much as the presence of some degree of continuous emoting is virtually always associated with the conscious state. A related finding occurs regularly during the natural experiment on consciousness we call sleep. Deep sleep is not accompanied by emotional expressions, but in dream sleep, during which consciousness returns in its odd way, emotional expressions are easily detectable in humans and in animals.

Finding parallel impairments of consciousness and emotion will seem all the more notable when we consider that patients in whom core consciousness is intact but extended consciousness is compromised have recognizably normal background and primary emotions. Emotions and core consciousness tend to go together, in the literal sense, by being present together or absent together.[6]

The lack of emotion is surprising given that, as we have seen, emotions can be triggered nonconsciously, from unattended thoughts or unknown dispositions, as well as from unperceivable aspects of our body states. The lack of emotion when core consciousness vanishes may be parsimoniously explained by suggesting that both emotions and core consciousness require, in part, the same neural substrates, and that strategically placed dysfunction compromises both kinds of processing. The shared substrates include the ensemble of neural structures which support the proto-self (to be described in chapter 5), the structures which both regulate and represent the body's internal states. I take the lack of emotion, from background emotion on up to higher levels of emotion, as a sign that important mechanisms of body regulation have been compromised. Core consciousness is functionally close to the disrupted mechanisms, interwoven with them, and thus compromised along with them. There is no such close functional relationship between emotional processing and extended consciousness. That is why, as noted in chapter 7, impairments of extended consciousness are not accompanied by a breakdown of emotion.

Subjects with normal consciousness can take stock of their emotions in the form of feelings, and those feelings, in turn, can generate a new melodic line of emotions that confers upon behavior the traits we so easily recognize as characteristic of sentient life. In the pathological condition, the suspension of the reverberating cycle of emotion-to-feeling-to-emotion robs behavior of a major telltale sign of sentience and generates in the observer the idea that something strange is going on in the mind of the subject observed. I would not be surprised to discover that the reason why we so confidently attribute consciousness to the minds of some animals, especially domestic animals, comes from the patently motivated flow of emotions they exhibit and from our automatic and reasonable assumption that such emotions are indeed caused by feelings that could only affect behavior in a sentient creature. I shall pursue this issue later.

ANOTHER IMPORTANT SOURCE of information regarding impaired consciousness comes from the study of patients with a condition known by the blanket term *akinetic mutism. Akinesia* is the technical term for lack of movement, usually due to an inability to initiate movement, although it often includes the slow execution of movement; *mutism,* as the word indicates, denotes an absence of speech. As usual, the terms are suggestive of what goes on externally, or does not, but miss the mark on the inside view. Internally, from all the available evidence, consciousness is severely diminished or even suspended altogether. The problem of so-called akinetic mutisms fascinated me for years and I spent many hours observing these patients, in their hospital beds or in my laboratory, studying their scans and electroencephalograms, and waiting patiently for their mutism to resolve so that I could perhaps talk to them. The story of one of my patients with this condition will give you an idea of what happens.

The stroke suffered by this patient, whom I will call L, produced damage to the internal and upper regions of the frontal lobe in both hemispheres. An area known as the cingulate cortex was damaged, along with nearby regions. She had suddenly become motionless and speechless, and, by and large, she was to remain motionless and speechless for the best part of the next six months. She would lie in bed, often with her eyes open but with a blank facial expression. On occasion she might catch an object in motion—me, for instance, moving around her bed—and track for a few instants, eyes and head moving along for a moment, but the quiet, nonfocused staring would be resumed rapidly. The term *neutral* helps convey the equanimity of her expression, but once you concentrated on her eyes, the word *vacuous* gets closer to the mark. She was there but not there.

Her body was no more animated than her face. She might make a normal movement with arm and hand, for instance, to pull her bed covers, but in general her limbs were in repose. Together, body and face never expressed any emotion of any kind, background, primary, or secondary, although there were plenty of inducers offered, day to day, in the attempts at focused conversations or just plain bedside banter of physicians, nurses, medical students, friends, and relatives. Emotional neutrality reigned supreme, meaning that not only was there no response to external inducers, but no response, either, to internal inducers, those that might be present in her thoughts but, as it turns out, obviously were not.

When asked about her situation she almost invariably remained silent, although, after much coaxing, she might say her name, just once, only to resume her silence. She had nothing to say about the events leading to her admission, nothing to comment on her past or present. She did not react to the presence of her relatives and friends any more than she did to her physician and nurses. Neither photographs nor songs, neither darkness nor bright light, neither claps of thunder nor the rustle of rain, could move her to react. She never became upset with my insistent and repetitive questioning, never showed a flicker of worry about herself or anything else.

Months later, as she emerged from this state of narrowed existence and gradually began to answer some questions, she would clarify the enigma of her state of mind. Contrary to what a casual observer might have thought, her mind had not been imprisoned in the jail of her immobility. Instead, it appeared that there had not been much mind at all, and nothing that would resemble core consciousness, let alone extended consciousness. The passivity in her face and body was the appropriate reflection of her lack of mental animation. She had no recall of any particular experience during her long period of silence; she had never felt fear; had never been anxious; had never wished to communicate. For the period that immediately preceded her first answers to me, a matter of perhaps a few days, she vaguely recalled that she was being asked questions, but she felt that she really had nothing to say, and again, that caused her no suffering. Nothing had forced her not to speak her mind.

Unlike the patients with locked-in syndrome (discussed in chapter 8), L seems not to have had any sense of self and surroundings, any sense of knowing, for most of her long waking slumber. Even during her slow awakening, it is likely that her sense of self was impaired. Unlike locked-in patients, but along with the epileptic patients described earlier and the patients described in the next section, L could have moved perfectly—limbs, eyes, speech apparatus—had she had a conscious mind to formulate a plan and command a movement. But she did not. Although some images were probably being formed—it is difficult to imagine how she could track an object or how she could pull her bed covers by touch, with precision, if she were relying exclusively on reflexes—it appears that she had not been producing differentiated thought, reasoning, or planning, and that there had been no emotional reaction to any mental content, either. That momentous set of defects had been translated externally in a neutral facial expression, a virtual suspension of body movement, and mutism. Again, emotion was missing.

IN SOME PATIENTS with advanced stages of Alzheimer's disease, consciousness is also impaired, and in a manner similar to the one just described for akinetic mutism. Early in the disease, memory loss dominates the picture and consciousness is intact, but as the ravages of Alzheimer's deepen, one often finds a progressive degradation of consciousness. Unfortunately, textbooks and lay descriptions of Alzheimer's emphasize the loss of memory and the early preservation of consciousness and often fail to mention this important aspect of the disease.

The decline first affects extended consciousness by narrowing its scope progressively to the point in which virtually all semblance of autobiographical self disappears. Eventually, it is the turn of core consciousness to be diminished to a degree in which even the simple sense of self is no longer present. Wakefulness is maintained and patients respond to people and objects in elementary fashion—a look or a touch, the holding of an object—but there is no sign that the responses issue from real knowing. In a matter of a few seconds, the continuity of the patients' attention is disrupted, and the lack of overall purpose becomes evident.

I have seen this disintegration occur in many Alzheimer's patients and never as painfully as in a dear friend who was also one of the notable philosophers of his generation and whose intellectual brilliance disguised his mental decline for all but those closest to him. On the last occasion I saw him, he uttered no word and gave no sign of recognizing me or his wife. His eyes, whose expression had been emptied out from within, would settle on a person or object for a few seconds, without any reaction ensuing in his face or body. No sign of emotion at all would ever arise, positive or negative. And yet he could make his wheelchair move, here and there about the room, somewhat unpredictably, for instance, to approach the large picture window and look out at nothing in particular.

Once, I saw him move close to the single, nearly empty bookcase in the room, reach for a shelf at about the level of the chair's armrest, and pick up a folded paper. It was a worn-out glossy print, 8 × 10, folded in four. He set it on his lap, slowly; he unfolded it, slowly; and he stared for a long time at the beautiful face in it, that of his smiling wife, now split in four quadrants by the deep creases in the countlessly folded paper. He looked but did not see. There was no glimmer of reaction, at any moment, no connection made between the portrait and its living model who was sitting across from him, only a few feet away; no connection made to me, either, who had actually made the photograph ten years before, at a time of shared joy. The folding and unfolding of the photograph had happened regularly, from earlier in the progress of the disease, when he still knew that something was amiss, perhaps as a desperate attempt to cling to the certainty of what once was. Now it had become an unconscious ritual, performed with the same slow pace, in the same silence, with the same lack of affective resonance. In the sadness of the moment I was happy that he no longer could know.

REFLECTION ON THESE instances of disturbed consciousness reveals the following facts:

First, there is a sharp separation between, on the one hand, wakefulness, low-level attention, and brief, adequate behaviors, which can survive the disturbance of consciousness, and, on the other, emotion, which is lost along with the sense of knowing and self. The defect of knowing and self and of recognizably motivated emotion goes hand in hand with defects in planning, in high-level attention, and in sustained and adequate behaviors. The decoupling of functions that we can observe in these cases exposes a layering of subcomponents which would have been difficult to notice, let alone tease apart, without the scalpel provided by neurological disease.

Second, for practical purposes we can classify the neurological examples of disrupted core consciousness as follows:

A. *Disruption of core consciousness with preserved wakefulness and preserved minimal attention/behavior.* The prime examples are akinetic mutisms and epileptic automatisms. Akinetic mutisms are caused by dysfunction in the cingulate cortex, in the basal forebrain, in the thalamus, and in the medial, peri-cingulate parietal cortex.

B. *Disruption of core consciousness with preserved wakefulness but defective minimal attention/behavior.* Absence seizures and persistent vegetative state are the prime examples. Absence seizures are related to dysfunction in the thalamus or in the anterior cingulate cortex.

Persistent vegetative state, which is often confused with coma, can be distinguished from coma in that vegetative patients have cycles of sleep and wakefulness as shown by the opening and closing of their eyes and, sometimes, by their EEG patterns. Persistent vegetative state is discussed in chapter 8. It is frequently caused by dysfunction in a particular set of structures in the upper brain stem, hypothalamus, or thalamus.

C. *Disruption of core consciousness accompanied by disruption of wakefulness.* The examples are coma, the transient loss of consciousness caused by head injury or fainting, deep (dreamless) sleep, and deep anesthesia. Relevant aspects of coma are discussed in chapter 8, but we note that the typical site of dysfunction is in structures of the upper brain stem, hypothalamus, and thalamus. The control of sleep and wakefulness resides in the same general region, and the action of several anesthetics is known to take place in that region, too.

Third, as will become clear when we discuss the neuroanatomical correlates of consciousness (in chapters 6 and 8), nearly all the sites of brain damage associated with a significant disruption of core consciousness share one important trait: they are located near the brain's midline, in fact, the left and right sides of these structures are like mirror images, looking at each other across the midline. At the level of the brain stem and diencephalon (the region that encompasses the thalamus and hypothalamus), the damaged sites are close to the long set of canals and ventricles that define the midline of the entire central nervous system. At cortical level, they are located in the medial (internal) surface of the brain. None of them can be seen when we inspect the lateral (external) surfaces of the brain, and all of them occupy an intriguingly "central" position. These structures are of old evolutionary vintage, they are present in numerous nonhuman species, and they mature early in individual human development.

Chapter Four

The Hint Half Hinted

LANGUAGE AND CONSCIOUSNESS

On several occasions when I was in medical school and in neurology training, I remember asking some of the wisest people around me how we produced the conscious mind. Curiously, I always got the same answer: language did it. I was told that creatures without language were limited to their uncognizant existence but not we fortunate humans because language made us know. Consciousness was a verbal interpretation of ongoing mental processes. Language also gave us the requisite remove to look at things from a proper distance. The answer sounded too easy, far too simple for something which I then imagined unconquerably complex, and also quite implausible, given what I saw when I went to the zoo. I never believed it and I am glad I did not.

Language—that is, words and sentences—is a translation of something else, a conversion from nonlinguistic images which stand for entities, events, relationships, and inferences. If language operates for the self and for consciousness in the same way that it operates for everything else, that is, by symbolizing in words and sentences what exists first in a nonverbal form, then there must be a nonverbal self and a nonverbal knowing for which the words "I" or "me" or the phrase "I know" are the appropriate translations, in any language. I believe it is legitimate to take the phrase "I know" and deduce from it the presence of a nonverbal image of knowing centered on a self that precedes and motivates that verbal phrase.

The idea that self and consciousness would emerge *after* language, and would be a direct construction of language, is not likely to be correct. Language does not come out of nothing. Language gives us names for things. If self and consciousness were born de novo from language, they would constitute the sole instance of words without an underlying concept.

Given our supreme language gift, most of the ingredients of consciousness, from objects to inferences, can be translated into language, and for us, at this point in the history of nature and the history of each individual, the basic process of consciousness is relentlessly translated by language, covered by it, if you will. Language is a major contributor to the high-level form of consciousness which we are using at this very moment, and which I call extended consciousness. Because of this, it does require a major effort to imagine what lies behind language, but the effort must be made.

If You Had That Much Money:
A Comment on Language and Consciousness

As I studied case after case of patients with severe language disorders caused by neurological diseases, I realized that no matter how much impairment of language there was, the patient's thought processes remained intact in their essentials, and, more importantly, the patient's consciousness of his or her situation seemed no different from mine. The contribution of language to the mind was, to say the least, astounding, but its contribution to core consciousness was nowhere to be found.

This should be no surprise when we consider where language

stands in the grand scheme of mental abilities. Is it plausible to think that language utterances could be created in individuals who had no sense of self, other, and surroundings?

In every instance I know, patients with major language impairments remain awake and attentive and can behave purposefully. More importantly, they are quite capable of signaling that they are experiencing a particular object, or detecting the humor or tragedy of a situation, or picturing an outcome that the observer anticipates. The signaling can be made via impoverished language or via a hand gesture, body movement, or facial expression, but it is there, promptly. Just as importantly, emotion is present in abundance in the form of background, primary, and secondary emotions, richly connected to the ongoing events, obviously motivated by them, recognizably comparable to what our own emotion would be in comparable situations.

The best evidence, in this regard, comes from patients with what is know as global aphasia. This is a major breakdown of *all* language faculties. Patients are unable to comprehend language whether auditorily or visually. In other words, they understand no speech when spoken to and they cannot read a single word or letter; they have no ability to produce speech beyond stereotypical words, largely curse words; they cannot even repeat a word or sound if you ask them to. There is no evidence that, in their awake and attentive minds, any words or sentences are being formed. On the contrary, there is much to suggest that theirs is a wordless thought process.

Yet, while it is out of the question to maintain a normal conversation with a global aphasic, it *is* possible to communicate, richly and humanly, if only you have the patience to accommodate to the limited and improvised vocabulary of nonlinguistic signs the patient may develop. As you familiarize yourself with the tools at the patient's disposal, it will never even cross your mind to ask if that human being is or is not conscious. In terms of core consciousness, that human being is no different from you and me, despite the inability to translate thought into language and vice versa.

Now let me play devil's advocate and see where I land. In patients with global aphasia, the damage destroys a large sector of the left cerebral hemisphere but does not destroy it completely. Patients with global aphasia have damage to both famous language areas, Broca's and Wernicke's, in the frontal and temporal lobe of the left hemisphere; they usually have extensive damage to regions of the frontal, parietal, and temporal cortices in between Broca's and Wernicke's areas, and damage to a vast amount of white matter underneath these cortices and even to gray matter in the basal ganglia of the left hemisphere. It might be argued by the skeptics, however, that even in the worst cases of global aphasia, there are still some portions of the left hemisphere that remain intact in the prefrontal and occipital regions. Might it be the case that such regions, while not able to permit proper speech, retain some of the "language-related" abilities that are necessary for "language-caused" consciousness to emerge?

This possibility can be addressed directly by studying the behavior of patients who underwent radical excisions of the entire left hemisphere for the treatment of certain brain tumors. This kind of operation, which is no longer in use but was once practiced as a last resort to manage the situation of patients with malignant and rapidly fatal brain tumors, called for the removal of the entire hemisphere within which the tumor was harbored, i.e., no cerebral cortex was left behind, not even in the areas that the skeptics in my thought experiment might invoke. Left hemispherectomies, as one might expect, were devastating from the point of view of language, resulting in nothing short of the most severe kind of global aphasia. But I have a vivid image of some of those patients, and I will tell you of one patient in particular, named Earl, who was studied by Norman Geschwind in the mid 1960s.

I can assure you that the intactness of Earl's core consciousness was not questioned at the time, nor would it be questioned today. Although Earl's language production was virtually confined to a few expletives, it was apparent that he used them with perfect intention to indicate what he thought of questions, of parts of the examination,

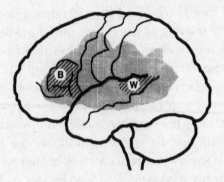

Figure 4.1. Minimal extent of damage in the left cerebral hemisphere of a typical patient with global aphasia. Broca's and Wernicke's areas are destroyed and so are several other areas involved in language processing, cortically and subcortically.

and of his own frustratingly limited abilities. Earl was not just awake and attentive, he also produced behavior appropriate to the wretched lot life accorded to him. He was not just producing thoughtless, consciousnessless reflexes. He *attempted* to respond to the questions one posed, sometimes by using gestures, and there were thoughtful delays between figuring out what on earth the examiner's pantomimes meant and concluding that he could not produce an answer. Sometimes he would respond with a movement of the head or a facial expression. Sometimes the frustration would be transmitted in a telling hand gesture filled with resignation. The melody of his emotions was finely tuned to the moment.

Language hardly needs consciousness as one more among the important abilities that humans should thank it for. The glories of language lie elsewhere, in the ability to translate, with precision, thoughts into words and sentences, and words and sentences into thoughts; in the ability to classify knowledge rapidly and economically under the protective umbrella of a word; and in the ability to express imaginary constructions or distant abstractions with an efficient simple word. But none of these remarkable abilities—which have allowed the human mind to grow in knowledge, intelligence, and creativity and have strengthened the sophisticated forms of *extended* consciousness that are ours today—have anything to do with manufacturing core consciousness, any more than they have to do with manufacturing emotion or perception.

I always remember fondly a sweet grandmother whose stroke had left her with a severe aphasia and who, with the willpower and intelligence permitted by her conscious mind, was determined to overcome her defects. She did improve remarkably, but her language remained a pale shadow of what it once was, and not everyone would sign up to hear her speak. One day I was checking on her ability to produce names of unique individuals and I was showing her a series of celebrity photos and asking her for the names of each person. We came to a glamour shot of Nancy Reagan—this being the splurgy 1980s—and there was Mrs. Reagan, clad in something shiny and silvery, her hair shimmering; her gaze sparkling, cast upward on Ronnie. My lovely patient's wrinkled face grew somber and although she could not come up with Nancy Reagan's name, she uttered: "If *you* had that much money, *I* would be like that *either*." How movingly conscious of her! She had instantly seized on the several layers of meaning implied by this iconic picture. But although she had managed to select a few words correctly, and had even come up with a correct conditional frame for her utterance, she was not even able to find consistently the correct pronoun to denote herself—language could not provide a stable translation for her *self* or for another. Her language could no longer match the sophistication of her thought process, and yet, what a rich autobiographical self was still available to her.

MEMORY AND CONSCIOUSNESS

Just as language can be exonerated from any role in the creation of core consciousness, so can conventional memory. Core consciousness is not founded on extensive memory. It is not founded on working memory, either, which is, however, required for extended consciousness. In terms of memory, all that core consciousness requires is a very brief, short-term memory. We do not require access to vast stores of past personal memories to have core consciousness, although such vast autobiographical reservoirs contribute to the advanced levels of consciousness I designate as extended consciousness. My views on this issue were shaped by the investigation of patients with severe disorders of learning and memory, the so-called amnesias. I will illustrate my point with a patient of mine, David, perhaps the most profoundly amnesic patient on record, and whom I have been studying for over twenty years. I talked about David when I reported the results of the good-guy/bad-guy experiment, and here he is now, in person.

Nothing Comes to Mind

My friend David has just arrived. I greet him with a hug and a smile, and he returns the gesture. I am delighted to see him and he is delighted to see me. It is all so natural that I cannot tell you who smiled first or who first moved toward the other. It does not matter. David and I are pleased to be here. We sit and begin to talk as old friends do. I offer David coffee and pour some for me. If you were watching innocently on the other side of the glass, you would have nothing unusual to report.

But the scene is about to change. Breaking from the convention of pleasant conversations among friends, I now turn to David and ask him who I am. Unfazed, David says that I am his friend. Unfazed, I say, "Of course. But David, who am I really, what is my name?"

"Well, I do not know, I can't think of it now, I just can't."

"But, David, please try to remember my name."

And David does answer then. "You are my cousin George."

"George who, David, please, cousin George who?"

"Cousin George McKenzie," says David, with an assertive voice but a quizzical, fleeting furrowing of his brow.

Everyone knows that I am not George McKenzie and that I am not David's cousin—everyone but David, obviously. Appearances to the contrary, David does not know who I am. He does not know what I do, he does not know if he has seen me before or not, he does not know when he last saw me, and he does not know my name. Nor does he know the name of the city he is in, or the name of the street, or the name of the building. He does not know what time it is, either, although when I ask him the time he promptly looks at his watch and says, correctly, a quarter to three. When I ask him about the date he looks again at the watch and says, again correctly, that it is the sixth of the month. His watch has a prominent window for the day of the month but not for the month.

"Perfect, David, perfect, but what month please?"

To which he says, looking uneasily about the room and glancing at the tightly drawn curtains on the window, "Well, February or March, I believe; it's been rather cold"; and without missing a beat, halfway through the last sentence he has gotten up, walked to the window, drawn the curtains about, and exclaimed, "Oh, heavens no! It must be June or July; it's really summerish weather."

"Indeed it is," I say. "It is June and it is about ninety outside."

To which David retorts, "Ninety degrees above zero? My, how wonderful, we should go outside."

David returns to his chair and we resume our conversation. If I stay away from specifics of people, places, events, or times, the conversation returns to normal. David knows his way around in a nonspecific world. His words are well chosen; the speech is melodic; the prosody rich with the emotions appropriate to the moment; and his facial expressions, his hand and arm gestures, and the body posture he assumes as he relaxes in the chair are precisely as you would expect for the situation. David's background emotions flow like a large, wide river. But the spontaneous content of David's conversation is generic, and whenever he is asked to produce any nongeneric detail, he often declines to do so and confesses, quite candidly, that nothing comes to mind. Pressed to venture the specific description of an event, or to place it in time, or to offer the name of a unique person, he will throw caution to the winds and produce a fable.

My old friend David has one of the most profound memory impairments ever recorded in any human being. David's memory was entirely normal until the day he was struck by a severe encephalitis. In David's case, this infectious disease of the brain tissue was caused by a virus, the herpes simplex virus type I. Most of us carry the virus, but only a vanishingly small number of us will ever have encephalitis caused by it. No one knows why the virus suddenly behaves aggressively in the unfortunate few.

David was forty-six at the time he developed encephalitis. The disease caused major damage in selected regions of David's brain, namely, in the left and right temporal lobes. Once the disease process was over, in a matter of weeks, it became clear that David was unable to learn any new facts. He simply could not learn any new item. It made no difference whether he encountered a new person or a new landscape, whether he witnessed a new event or was given a new word to remember, he just would not retain any fact in memory. His memory was limited to a window of time of less than one minute. During that brief period his memory for new facts was normal. If I were to introduce myself to him, leave the room, and come back within, say, twenty seconds, to ask him who I was, he would promptly say my name and say that yes, he had just met me, that I had disappeared and had now returned. But if, instead, I were to return three minutes later, David would not have the faintest idea of who I was. And if I pressed him on it, I would become *anybody,* perhaps cousin George McKenzie.

In his profound inability to learn new facts, David was similar to patient HM, first studied in detail by the psychologist Brenda Milner. HM has been unable to learn any new facts since the mid 1950s (curiously, he is about the same age as David). But David's memory defect is more extensive than HM's because not only is he unable to learn new facts, but he is also unable to recall many old facts. The recall of virtually any unique thing, individual, or event, from his entire life, is denied to him. His memory loss goes almost all the way to the cradle.

There are few exceptions to this ravage. He does know his name and the names of his wife, children, and close relatives. He does not recall what they look like or what their voices sound like. Accordingly, he cannot recognize any of them in photographs, old or recent, and he does not recognize them in person. In fact, he fails to recognize most photos of himself, the exception being some of his photos as a young man. The reason why David and HM are both so similar in their inability to learn new facts, and so different in their ability to recall old facts, is that they share one site of damage, the hippocampus, but do not share another site of damage that is only compromised in David—the cortices in the remainder of the temporal lobe, especially those in the inferotemporal and polar region.

David knows his former professional occupation and the name of the city in which he lived most of his life, but he cannot picture the place and he cannot recognize photographs of his former houses, of the cars he owned, of the pets he loved, or of personal artifacts that were dear to him. Nothing specific comes to mind when he is asked about those unique items, and what comes to mind when he is shown photos of the items or the items themselves is the knowledge of the item as a member of a conceptual category. Shown a picture of his son, age fourteen, he says that it shows a young man with a nice smile, probably going to high school, but he has no idea that it is his own son. All that he remembers, as shown in the conversation above, are the generics of most everything in the world around him. He knows what a city is, and a street, and a building, and how a hospital differs from a hotel. He knows what different kinds of furniture, or clothes, or means of transportation are available. He also knows the different kinds of actions that things or living beings can perform, and he knows the general plot line of the events that most commonly involve such things or living beings. But when you realize that he has lost the ability to access the unique facts that he learned until age forty-six, and that he has not been able to acquire any new facts since then, you take stock of the magnitude of this impairment. So profound is the impairment that you may well wonder what the mind is like inside such a person. Is David a zombie, the kind of being some philosophers have created in their thought experiments? More to our point: Is David conscious?

David's Consciousness

David fares perfectly well on the core consciousness checklist. To begin with David has wakefulness. In the traditional wording of neurologists he is "awake and alert." We know, by the way, that his circadian rhythms are normal, that he sleeps normally, and that he spends the expected part of his sleeping time in REM sleep, the rapid-eye-movement period of sleep, during which dreams occur. There is also no question that David behaves attentively toward the stimuli we present to him. Whether he is asked to listen to a sentence or a piece of music, whether we show him a picture or a film, he attends to the stimulus as you or I would, sometimes with great enthusiasm, sometimes less so, but always adequately enough to process the stimulus, create an impression of it, and be ready to answer a question about it. His attention can be focused and sustained over substantial periods of time, long periods, in fact, provided the stimulus or the situation engage his interest. For instance, he can play a whole set of checkers—and win!—although he does not even know the name of the game and would not be able to articulate a single rule for it or indicate when he last played it. Background emotions flow continuously and so do many, though not all, primary and secondary emotions. His joy at winning the game is a delight to behold; the affective modulation of his voice as the game approaches its decision point is a primer of human emotion. Finally, his spontaneous behavior is purposeful—he will look appropriately for a good chair to sit in, for food and drink to consume, for a television screen or a window from which he can watch the world. Left to his own devices, he sustains purposeful behavior relative to the context he is in for many minutes or hours, provided that what he is doing is engaging.

The distinction between David and the patients I described earlier is quite clear. Patients with epileptic automatisms are also awake, but their attention span is most brief, is not sustained over an object, and stays on the object only for the time necessary to create an image and prompt the next behavior. The behavior of patients with automatism is purposeful only within each action (drinking from a glass) or for a few consecutive actions (getting up and walking out) but there is no continuity of purpose. The behaviors are not appropriate to the overall context of a situation.

On the grounds that normal wakefulness, attention, and purposeful behavior are present, users of an external definition of consciousness would conclude that David has normal consciousness. I would agree, of course, and to help the diagnosis of the externalists I would add that David is quite conscious of the relation between himself and his surroundings, as is clearly indicated by his report of personal reactions to the things and events around him. I cannot jump into his mind and take a look, but I can analyze his ever-present commentaries on the world he is experiencing—"Oh! This is terrific"; "I like this one"; "It's nice to be sitting here watching pictures with you guys"; "Gosh, how terrible"; "Tastes delicious to me; this is my favorite kind"; "I don't think it is nice to say those things in public." It is legitimate to deduce that since we are organisms of the same species and since this commentary is not formally different from the one that we would make in similar circumstances, it originates in a mental state that is formally comparable to that in which we would produce such judgments. When almost nothing comes to mind, David's sense of self still does.

Within the temporal window of his short-term memory—which lasts about forty-five seconds—there is ample time to generate core consciousness about a slew of items. There is evidence that the images David forms in the varied sensory modalities—vision, hearing, touch—are formed in the perspective of his organism. It is patently obvious that he treats those images as his, not as somebody else's. And it is easily observable that he can act on the basis of these images and reports intentions to act that are closely coupled to the content of the images. In conclusion, David is not a zombie. In terms of core consciousness, David is as conscious as you or I.

It goes without saying that David's mind is not entirely like yours or mine and it is important to describe what is missing. His mind is like ours in the sense that it has images of varied sensory modalities, that those images occur in coordinated and logically interconnected sets, that those sets change over time in a forward direction, and that new sets succeed the preceding ones. David has a stream of such image sets, the kind of process that Shakespeare and Joyce converted in literary form in their soliloquies, and that William James named a *stream of consciousness*. But the *content* in the images within David's stream of consciousness is a different matter. We know for certain that his images embody the general rather than the particular — general knowledge about the stimuli we show him and general knowledge about his person, about his body, about his current physical and mental states, about his likes and dislikes. Unlike us, David can never conjure up the specifics of unique things, persons, places, or events. Whereas you or I will inevitably mix, at every turn, images of general knowledge with images of unique knowledge, David is obliged to stay within the general. David's mind differs from ours in the specificity of its contents. I suspect it also differs in the quantity of images. By being limited to generic contents, David's mind may well process in each unit of time a smaller number of images than you or I do.

The sheer lack of specific content does compromise his ability to relate the apprehension of a given object to the comprehensive sweep of his historical person. He can sense the factual meaning of an object and develop a feeling of pleasure for it, but he cannot articulate how he developed the factual meaning or the feeling, he cannot recall which specific instances in his autobiography may have led to the images he conjures up. Nor can he articulate how that object does or does not relate to his anticipated future, for the simple reason that David has no memory of a planned, potential future, as you and I have. David has not been able to plan ahead because planning ahead requires the intelligent manipulation of specific images of the past and David cannot evoke any specific images. Everything indicates that he has a normal sense of self, in the here and now, but his autobiographical memory has been reduced to a skeleton, and thus the autobiographical self that can be constructed at any moment is severely impoverished.

As a result of this paucity of specifics, David's extended consciousness is impaired. It is possible that if he were able to conjure up the specific contents he no longer holds in his autobiographical memory, some of the mechanisms which permit extended consciousness might actually be in place. There is no evidence that he lacks the capacity to produce several mental images simultaneously or to hold in mind different images of different sensory modalities, a capacity enabled by working memory and essential for extended consciousness. For example, he can carry out tasks that require conjunctions of color, shape, and size without difficulty.

Because David lacks the specifics required to define unique items, he also lacks the aspects of extended consciousness concerned with social cognition and behavior. High-level awareness of social situa-

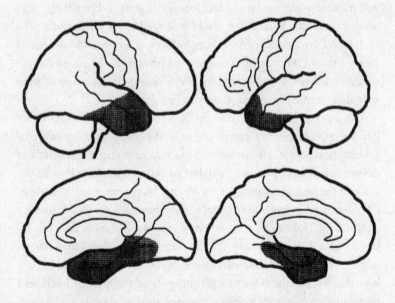

Figure 4.2. The extent of temporal lobe damage in patient David. The damage destroyed large sections of the temporal lobes, including the hippocampus, in both the left and right hemisphere. The learning of any new facts and the recall of old facts are severely impaired.

tions is built on a vast knowledge of specific social situations and David cannot evoke such knowledge. He observes a good number of social conventions as shown in the polite manner with which he greets others, takes turns in conversation, or walks about in a street or hallway. He also has a notion of what humane and kind behavior is like. But the comprehensive knowledge of the operations of a social collective eludes him.

David provides evidence to support two conclusions. The first is that factual knowledge at a unique and specific level is not a prerequisite for core consciousness. The second: David has extensive damage to both temporal regions, including the hippocampus, the medial cortices overlying it, the polar temporal region, a sizable sector of the lateral and inferior temporal regions, and the amygdala. We thus learn that core consciousness cannot depend on those vast brain regions at all.

ROUNDING UP SOME FACTS

A number of preliminary facts can be culled from this brief survey of conditions under which consciousness can be either impaired or left intact.

First, consciousness is not a monolith. It is reasonable to distinguish kinds of consciousness—there is at least one natural break between the simple, foundational kind and the complex, extended kind—and it is also reasonable to distinguish levels or grades within extended consciousness. The results of neurological disease validate the distinction between core consciousness and extended consciousness. The foundational kind of consciousness, core consciousness, is disrupted in akinetic mutisms, absence seizures, and epileptic automatisms, persistent vegetative state, coma, deep sleep (dreamless), and deep anesthesia. In keeping with the foundational nature of core consciousness, when core consciousness fails, extended consciousness fails as well. On the other hand, when extended consciousness is disrupted, as exemplified by patients with profound disturbances of autobiographical memory, core consciousness remains intact. (Extended consciousness and its disorders are addressed in chapter 7.)

Second, it is possible to separate consciousness in general from functions such as wakefulness, low-level attention, working memory, conventional memory, language, and reasoning. Core consciousness is not the same as wakefulness or low-level attention, although it requires both to operate normally. As we have seen, patients with absence seizures or automatisms or akinetic mutism are technically awake but not conscious. On the other hand, patients who lose wakefulness (the partial exception of REM sleep aside) can no longer be conscious.

Core consciousness is also not the same as holding an image over time, a process known as working memory—the sense of self and of knowing is so brief and so abundantly produced that there is no need to hold it over time in order for it to be effective. On the other hand, working memory is vital for the process of extended consciousness.

As we have seen, core consciousness does not depend on making a stable memory of an image or recalling it, that is, it does not depend on the processes of conventional learning and memory; core consciousness is not based on language; lastly, core consciousness is not equal to manipulating an image intelligently in processes such as planning, problem solving, and creativity. Patients with profound defects of reasoning and planning exhibit perfectly normal core consciousness although the top reaches of extended consciousness are then defective. (See *Descartes' Error.*)

All of these different aspects of cognition—wakefulness, image making, attention, working memory, conventional memory, language, intelligence—can be separated by appropriate analysis and investigated separately in spite of the fact that they operate together, in perfect concert with consciousness, as a most harmonious and virtuoso ensemble.

Third, emotion and core consciousness are clearly associated. Patients whose core consciousness is impaired do not reveal emotion by facial expression, body expression, or vocalization. The entire range of emotion, from background emotions to secondary emotions, is usually missing in these patients.[1] By contrast, as will be seen when we discuss extended consciousness (chapter 7), patients with preserved core consciousness but impaired extended consciousness have normal background and primary emotions. This association suggests, in the very least, that some of the neural devices on which both emotion and core consciousness depend are located within the same region. It is also plausible, however, that the connection between emotion and core consciousness goes beyond a mere contiguity of the neural devices on which they depend.

Fourth, disturbances of core consciousness target the entire realm of mental activity as well as the full range of sensory modalities. In patients with disturbed core consciousness, from those with coma and persistent vegetative state to those with epileptic automatisms, akinetic mutisms, and absence seizures, the impairment of core consciousness leaves no island of preserved consciousness. The impairment extends to all sensory modalities. Core consciousness serves the entire compass of thoughts that can be made conscious, the full scope of things to be known. Core consciousness is a central resource.

By contrast, as discussed in the next chapter, the impairment of image making within one sensory modality, e.g., visual or auditory, only compromises the conscious appreciation of one aspect of an object—the visual or the auditory—but not core consciousness in general and not even consciousness of the same object via a different sensory channel, e.g., olfactory or tactile. Naturally, an impairment of *all* image-making capability abolishes consciousness altogether because consciousness operates on images.

The above observations are not compatible with the idea that consciousness is broken down by sensory sector. There are conditions in which brain damage prevents patients from processing images of a certain kind, for instance, visual or auditory. In such cases, the sensory processing for that modality may be lost in its near entirety, as in cortical blindness, or one aspect of the modality may be lost, as in the loss of color processing known as achromatopsia, or a substantial part of a process may be disrupted, as when patients become unable to recognize familiar faces in the condition known as prosopagnosia. In my framework, the patients so affected have a disturbance of the "something-to-be-known." But they have normal core consciousness for all the images formed in other sensory modalities, and, no less importantly, they have normal core consciousness for the specific stimuli they fail to process normally. In other words, patients who cannot recognize a previously familiar face have normal core consciousness for the stimulus that confronts them, are fully aware that they do not know the face even if they should. They know that, in fact, it is a human face and that it is their sense of self in the act of knowing that is failing to know. Those patients have normal core consciousness, and a normal extended consciousness outside of the island of defective knowledge. Their circumscribed plight underscores the fact that core consciousness, and its resulting sense of self, is a central resource. These observations also raise questions about attempts to understand consciousness *comprehensively* within the domain of a single sensory modality, such as vision, without appealing to the notion of the overall organism which consciousness is serving. Those attempts may contribute to the elucidation of the first of the two problems of consciousness outlined in chapter 1—the problem of the movie-in-the-brain—but do not address problem number two—the problem of the sense of self in the act of knowing.[2]

The fact that core consciousness is separable from other cognitive processes does not mean that consciousness does not have an influence on them. On the contrary, as explained in chapter 6, core consciousness has a major influence on those other cognitive processes. Core consciousness focuses and enhances attention and working memory; core consciousness favors establishment of memories; core consciousness is indispensable for the normal operations of language; and core consciousness enlarges the scope of the intelligent manipulations we call planning, problem solving, and creativity.

In conclusion, individuals such as we are, endowed with extensive memory and intelligence, can manipulate facts logically, with or without the help of language, and produce inferences from those facts. But I am proposing that core consciousness can be distinguished from the inferences that we may draw regarding the contents of core consciousness. We can infer that the thoughts in our minds are created in our individual perspective; that we own them; that we can act on them; that the apparent protagonist of the relationship with the object is our organism. As I see it, however, core consciousness begins before those inferences: *it is the very evidence, the unvarnished sense of our individual organism in the act of knowing.*

All the cognitive properties discussed above have been potentiated by core consciousness and have, in turn, helped build extended consciousness on the foundation of core consciousness. The umbilical cord has never been severed, however. Behind extended consciousness, at each and every moment, lies the pulse of core consciousness. This may sound surprising, but it should not be. We still need digestion in order to enjoy Bach.

THE HINT HALF HINTED

It is time to say a bit more about core consciousness, now that we have discussed the circumstances in which it can either disappear or be remarkably preserved in spite of other important cognitive disturbances being present.

In the opening chapter of this book, I suggested that core consciousness includes an inner sense based on images. I also suggested that the particular images are those of a feeling. That inner sense conveys a powerful nonverbal message regarding the relationship between the organism and the object: that there is an individual subject in the relationship, a transiently constructed entity to which the knowledge of the moment is seemingly attributed. Implicit in the message is the idea that the images of any given object that are now being processed are formed in our individual perspective, that we are the owners of the thought process, and that we can act on the contents of the thought process. The tail end of the core consciousness process includes the enhancement of the object that initiated it, so that the object becomes salient as part of the relationship it holds with the knower organism.

The view of consciousness I adopt here connects historically with those expressed by thinkers as diverse as Locke, Brentano, Kant, Freud, and William James. They believed as I do that consciousness is "an inner sense." Curiously, the "inner sense" view is no longer mainstream in consciousness studies.[3] In the view I adopt here, consciousness also conforms to the fundamental properties William James outlined for it: It is selective; it is continuous; it pertains to objects other than itself; it is personal. James did not make a distinction between core and extended kinds of consciousness, but that turns out not to pose a problem because the properties he proposed easily apply to both kinds of consciousness.[4]

Core consciousness is generated in pulselike fashion, for each content of which we are to be conscious. It is the knowledge that materializes when you confront an object, construct a neural pattern for it, and discover automatically that the now-salient image of the object is formed in your perspective, belongs to you, and that you can even act on it. You come by this knowledge, this discovery as I prefer to call it, instantly: there is no noticeable process of inference, no out-in-the-daylight logical process that leads you there, and no words at all—there is the image of the thing and, right next to it, is the sensing of its possession by you.

What you do not ever come to know directly is the mechanism behind the discovery, the steps that need to take place behind the seemingly open stage of your mind in order for core consciousness of an object's image to arise and make the image yours. Altogether, the steps behind the stage take time, time being of the essence to establish the causal link between the image of an object and its possession by you. The time elapsed is minuscule if measured by a fine stopwatch, but it is actually quite extensive if you think of it from the perspective of the neurons which make it all possible and whose units of time are so much smaller than that of your conscious mind—neurons get excited and fire themselves away in just a few milliseconds, while the events of which we are conscious in our minds occur in the order of many tens, hundreds, and thousands of milliseconds. By the time you get "delivery" of consciousness for a given object, things have been ticking away in the machinery of your brain for what would seem like an eternity to a molecule—if molecules could think. We are always hopelessly late for consciousness and because we all suffer from the same tardiness no one notices it. The idea that consciousness is tardy, relative to the entity that initiates the process of consciousness, is supported by Benjamin Libet's pioneering experiments on the time it takes for a stimulus to be made conscious. We are probably late for consciousness by about five hundred milliseconds.[5] It is curious, of course, that we can position our mental self between cellular time, on the one hand, and the time evolution has taken to bring us here, on the other, and humbling, for sure, that we cannot imagine properly either of those faraway time scales.

As you look at this page and see these words, whether you wish for it or not, automatically and relentlessly, you sense that *you* are doing the reading. I am not doing it, nor is anyone else. You are. You sense that the objects you are perceiving now—the book, the room around you, the street outside the window—are being apprehended in your perspective, and that the thoughts formed in your mind are yours, not anyone else's. You also sense that you can act on the scene if you so wish—stop reading, start reflecting, get up and take a walk. *Consciousness* is the umbrella term for the mental phenomena that permit the strange confection of you as observer or knower of the things observed, of you as owner of thoughts formed in your perspective, of you as potential agent on the scene. Consciousness is a part of your mental process rather than external to it. Individual perspective, individual ownership of thought, and individual agency are the critical riches that core consciousness contributes to the mental process now unfolding in your organism. The essence of core consciousness is the very thought of you—the very feeling of you—as an individual being involved in the process of knowing of your own existence and of the existence of others. Never mind, for the moment, that knowing and self, which are real mental entities, will turn out to be, biologically speaking, perfectly real but quite different from what our intuitions might lead us to imagine.

You are reading this text and translating the meaning of its words in conceptual thought flow as you go along. The words and sentences on the page, which are the translation of my concepts, become translated in turn, in your mind, by nonverbal images. The collection of those images defines the concepts that were originally in my mind. But in parallel with perceiving the printed words and displaying the corresponding conceptual knowledge required to understand them, your mind also represents *you* doing the reading and the understanding, moment by moment. The full scope of your mind is not confined to images of what is being perceived externally or of what is recalled relative to what is perceived. It also includes you.

The images that constitute knowing and sense of self—the feelings of knowing—do not command center stage in your mind. They influence mind most powerfully and yet they generally remain to the side; they use discretion. More often than not, knowing and sense of self are in subtle rather than assertive mode. It is the destiny of subtle mental contents to be missed, and not just those that constitute knowing and self.

Consider your current task: the words on the page and the thoughts they engender require, in traditional psychological terms, a procedure called attention, something of a finite commodity when it comes to real-time mental processing. My words and your thoughts command nearly all the processing capacity that you have available. In all probability, you are not simultaneously attentive to all the images that you are currently evoking as you analyze this text, let alone attentive to other images that you are also evoking and are unrelated. Because of this, some of your thoughts are likely to gain salience while others recede from the mental foreground—for instance, the words on the page may blur or disappear altogether, for a few moments, as you consider other images in your thought process. Discretion and subtleness are thus not unfairly directed at the signifier of you. They are a standard mode of operation for the mind.

A considerable number of the images formed on *any* topic go unnoticed or barely noticed at one time or another. Just a few minutes ago, the following happened: I was coming up to my study with a book in my left hand and a cup of coffee in my right. Earlier, midway up the stairs, I had left two pens on a step. As I climbed the stairs, without noticing any thought on this matter whatsoever, smoothly and swiftly, I transferred the cup to my left hand, a skilled action that required a precise movement so as not to spill the coffee and that also entailed slipping the book under my left arm; I then proceeded to pick up the pens with the right hand. In retrospect, all of these actions, which are not routine in this setting and sequence, were occurring seamlessly and seemingly thoughtlessly. In fact, I only noticed that there was a "plan" behind these actions when I saw how my right hand had adopted the shape necessary for the prehension of the two pens given their spatial orientation. For a split second, turning the focus of my mind to what had just gone on rather than to the very moment, I could reconstruct a part of the sensory-motor process behind this trivial and yet complex event.

Only a fraction of what goes on mentally is really clean enough and well lit enough to be noticed, and yet it is there, not far at all, and perhaps available if only you try. Curiously, one's context does influence how much one notices in the fringes of the mind. Were I not preoccupied with the issue of the subtle presence of the core self, I probably would not have noticed this incident at all and would not have reflected on the wealth of mental detail that accompanied these unremarkable acts.[6]

If you were to argue that you never notice yourself knowing, I would say, pay closer attention and you will. I would also add that it is advantageous not to notice yourself knowing. Come to think of it, unless the particular purpose of the mental moment was to reflect on a particular state of your organism, there would be little point in allocating attention to the part of mental contents which constitute the you of the moment, no need to waste processing capacity on you alone. Just let you be.

The fact that the signifier of you can use discretion does not mean that the signifier is unimportant or dispensable. You can, to a certain extent, willfully control the activity of the more elaborate sense of you that I call the autobiographical self; you can allow it to dominate the panorama of your mind, or be minimal. But you cannot do much about the presence of the core you; you cannot make it vanish entirely—a substantial presence always remains and a good thing, too. As we have just seen, the removal of core consciousness, except for those situations in which it is caused by sleep or anesthesia, is a sign of disease. If the removal is only partial, it causes an anomalous state which others will easily recognize as abnormal but which you will not know about—when there is no knowing, you do not know. Importantly, removal of knowing and self without removal of wakefulness places the organism in grave danger—one is then capable of acting without knowing the consequences of one's acts. It is as if, without the sense of self in the act of knowing, the thoughts one generates go unclaimed because their rightful owner is missing. The self-impoverished organism is at a loss as to whom those thoughts belong.

PART III

A Biology for Knowing

Chapter Five

The Organism and the Object

THE BODY BEHIND THE SELF

Focusing the investigation of consciousness on the problems of self made the inquiry all the more interesting but not any more clear until I began seeing consciousness in terms of two players, the *organism* and the *object*, and in terms of the *relationships* those players hold. All of a sudden, consciousness consisted of constructing knowledge about two facts: that the organism is involved in relating to some object, and that the object in the relation is causing a change in the organism. As previously noted, elucidating the biology of consciousness became a matter of discovering how the brain can construct neural patterns that map each of the two players and the relationships they hold.

The problem of representing the object seems less enigmatic than that of representing the organism. Neuroscience has dedicated considerable effort to understanding the neural basis of object representation. Extensive studies of perception, learning and memory, and language have given us a workable idea of how the brain processes an object, in sensory and motor terms, and an idea of how knowledge about an object can be stored in memory, categorized in conceptual or linguistic terms, and retrieved in recall or recognition modes. The object is exhibited, in the form of neural patterns, in the sensory cortices appropriate for its nature. For example, in the case of the visual aspects of an object, the appropriate neural patterns are constructed in a variety of regions of the visual cortices, not just one or two but many, working in concerted fashion to map the varied aspects of the object in visual terms. We will return to the representation of the object later in the chapter.

On the side of the organism, however, matters are different. Although much has been known about how the organism is represented in the brain, the idea that such representations could be linked to the mind and to the notion of self has received little attention. The question of what might give the brain a natural means to generate the singular and stable reference we call self has remained unanswered. I have believed for quite some time that the answer lies in a particular set of representations of the organism and of its potential actions. In *Descartes' Error* I advanced the possibility that the part of the mind we call self was, biologically speaking, grounded on a collection of nonconscious neural patterns standing for the part of the organism we call the body proper.[1] This may sound terribly odd at first hearing, but perhaps it should appear plausible after my reasons are considered.

The Need for Stability

In thinking about the biological roots for the procession of self from the simple core self to the elaborate autobiographical self, I began by considering their shared characteristics. At the top of the list I placed stability, and here is why. In all the kinds of self we can consider one notion always commands center stage: the notion of a bounded, single individual that changes ever so gently across time but, somehow, seems to stay the same. In highlighting stability I do not mean to suggest that self, in whatever version, is an immutable cognitive or neural entity, but rather that it must possess a remarkable degree of structural invariance so that it can dispense continuity of reference across long periods of time. Continuity of reference is in effect what the self needs to offer.

Relative stability is required at all levels of processing, from the simplest to the most complex. Stability must be there when you relate to varied objects in space or when you consistently react emotionally in a certain way to certain situations. Stability is there, too, at the level of complicated ideas. When I say, "I have changed my mind about corporations," I indicate that I once held certain opinions about corporations that I no longer do. The contents of my mind which describe corporations now and my concept of their behavior now have changed, but my "self" has not, or at least not to the same degree that my ideas about corporations have. Relative stability supports continuity of reference and is thus a requisite for the self. Our search for a biological substrate for the self must identify structures capable of providing such stability.

As we look behind the notion of self, we find the notion of the singular individual. And as we look behind individual singularity, we find stability. The riddle of the biological roots of the self can be worded like this then: What is it that provides the mind with a spine, is single, and is same?

The Internal Milieu as a Precursor to the Self

Consciousness is an important property of living organisms and it may be helpful to include life in its discussion. Consciousness certainly appears to postdate both life and the basic devices that allow organisms to maintain life, and in all likelihood, consciousness has succeeded in evolution precisely because it supports life most beautifully.

One key to understanding living organisms, from those that are made up of one cell to those that are made up of billions of cells, is the definition of their boundary, the separation between what is *in* and what is *out*. The structure of the organism is inside the boundary and the life of the organism is defined by the maintenance of internal states within the boundary. Singular individuality depends on the boundary.

Through thick and thin, even when large variations occur in the environment that surrounds an organism, there is a dispositional arrangement available in the organism's structure that modifies the inner workings of the organism. The dispositional arrangement ensures that the environmental variations do not cause a correspondingly large and excessive variation of activity within. When variations that trespass into a dangerous range are about to occur, they can be averted by some preemptive action; and when dangerous variations have already occurred, they can still be corrected by some appropriate action.

The specifications for survival that I am describing here include: a boundary; an internal structure; a dispositional arrangement for the regulation of internal states that subsumes a mandate to maintain life; a narrow range of variability of internal states so that those states are relatively stable. Now consider these specifications. Am I describing *just* a list of specifications for the survival of a simple living organism, or could it be that I am also describing some of the biological antecedents of the sense of self—the sense of a single, bounded, living organism bent on maintaining stability to maintain its life? I would say that I might be describing either. It is intriguing to think that the constancy of the internal milieu is essential to maintain life *and* that it might be a blueprint and anchor for what will eventually become a self in the mind.

More on the Internal Milieu

A simple organism made up of one single cell, say, an amoeba, is not just alive but bent on staying alive. Being a brainless and mindless creature, an amoeba does not know of its own organism's intentions in the sense that we know of our equivalent intentions. But the form of an intention is there, nonetheless, expressed by the manner in which the little creature manages to keep the chemical profile of its internal milieu in balance while around it, in the environment external to it, all hell may be breaking loose.

What I am driving at is that the urge to stay alive is not a modern development. It is not a property of humans alone. In some fashion or other, from simple to complex, most living organisms exhibit it. What does vary is the degree to which organisms *know* about that urge. Few do. But the urge is still there whether organisms know of it or not. Thanks to consciousness, humans are keenly aware of it.

Life is carried out inside a boundary that defines a body. Life and the life urge exist inside a boundary, the selectively permeable wall that separates the internal environment from the external environment. The idea of organism revolves around the existence of that boundary. In a single cell, the boundary is called a membrane. In complex creatures, like us, it takes many forms—for instance, the skin that covers most of our bodies; the cornea that covers the part of the eyeball that admits light; the mucosae that cover the mouth. If there is no boundary, there is no body, and if there is no body, there is no organism. Life needs a boundary. I believe that minds and consciousness, when they eventually appeared in evolution, were first and foremost about life and the life urge within a boundary. To a great extent they still are.

Under the Microscope

Now, look inside the boundary of a single cell. You will find the cell's nucleus immersed in a rich bath called the cytoplasm. Also immersed in the cytoplasm are the organelles, subdepartments of a cell such as the mitochondria and the microtubules. Life goes on only as long as the chemical profile of the bath operates within a certain range of possible variation. Life stops when the variation of a set of chemical parameters goes beyond or beneath certain values. In a curious way, life consists of continuous variation but only if the range of variation is contained within certain limits. If you were to look closely inside the boundary, life consists of one big change after another, an agitated sea with one high swelling wave following another. But if you look from a distance, the changes smooth out, like when a choppy ocean becomes a glassy surface seen from a plane high in the sky. And if you remove yourself even farther and look simultaneously at the whole cell and at

its environment, you will see that against the upheavals of the surroundings, life inside the cell is now largely stability and sameness.

The job of reining in the amplitude of changes, of keeping the inside in check against the odds from the outside, is a big task. It goes on
continuously, enabled by sharply targeted command and control
functions distributed throughout the cell nucleus, the organelles, and
the cytoplasm. In 1865, the French biologist, Claude Bernard, gave a
name to the environment inside an organism: the *internal milieu*. The
term has stuck, with its Gallic flavor, and no one ever uses "internal
environment" as a possible translation. Claude Bernard noted that
the chemical profile of the fluid within which cells live is usually quite
stable, varying only within narrow ranges, regardless of how large the
changes are in the environment surrounding the organism. His powerful insight was that in order for independent life to continue, the internal milieu had to be stable. In the earlier part of the twentieth
century, W. B. Cannon would carry these ideas forward by writing
about a biological function he named *homeostasis* and described as "the
coordinated physiological reactions which maintain most of the
steady states of the body . . . and which are so peculiar to the living
organism."[2]

The unwitting and unconscious urge to stay alive betrays itself inside a simple cell in a complicated operation that requires "sensing"
the state of the chemical profile inside the boundary, and that requires
unwitting, "unconscious knowledge" of what to do, chemically
speaking, when the sensing reveals too little or too much of some ingredient at some place or time within the cell. To put it in other
words: it requires something not unlike perception in order to sense
imbalance; it requires something not unlike implicit memory, in the
form of dispositions for action, in order to hold its technical know-
how; and it requires something not unlike a skill to perform a preemptive or corrective action. If all this sounds to you like the
description of important functions of our brain, you are correct. The
fact is, however, that I am not talking about a brain, because there is
no nervous system inside the little cell. Moreover, this brainlike mech-

anism that is not really a brain cannot be the result of nature copying the properties of a brain. On the contrary, sensing environmental conditions, holding know-how in dispositions, and acting on the basis of those dispositions were already present in single-cell creatures before they were part of any multicellular organisms, let alone multicellular organisms *with* brains.

Life and the life urge inside the boundary that circumscribes an organism precede the appearance of nervous systems, of brains. But when brains appear on the scene, they are still about life, and they do preserve and expand the ability to sense the internal state, to hold know-how in dispositions, and to use those dispositions to respond to changes in the environment that surrounds brains. Brains permit the life urge to be regulated ever so effectively and, at some point in evolution, knowingly.

Managing Life

The management of life poses different problems for different organisms in different environments. Simple organisms in hospitable environments may require little knowledge and no planning at all in order to respond adequately and maintain life. All that may be required is a few sensing devices, a stock of dispositions to respond according to what is sensed, and some means to carry out the action selected as response. By contrast, complex organisms placed in complex environments require large repertoires of knowledge, the possibility of choosing among many available responses, the ability to construct novel combinations of response, and the ability to plan ahead so as to avoid disadvantageous situations and instead propitiate favorable ones.

The machinery needed to perform these demanding tasks is complicated and requires a nervous system. It requires a vast stock of dispositions, a substantial part of which must be provided by the genome and be innate, although some dispositions can be modified by learning and additional stocks of dispositions can be acquired through experience. The control of the emotions, which we discussed earlier, is part of this dispositional stock. Several types of sensors are also required; these sensors must be capable of detecting varied signals from the environments external to the brain (the body) and external to the body (the outside world). Eventually, the management of life also requires a means of responding not just with actions carried out by muscles but also with images capable of depicting the internal states of the organism, entities, actions, and relationships.

Managing the life of a complex organism in a complex and not necessarily favorable environment thus requires more innate know-how, more sensing possibilities, and more variety of possible responses than a simple organism would need in a simple environment. But the issue is not just quantity. A new approach is needed and nature has permitted it by developing two anatomical and functional arrangements. The first consists of connecting the brain structures necessary to manage different aspects of the organism's life to an integrated but multiple-component system. An analogy from engineering would be the assembly of interconnected control panels. In biological terms, these panels are not myths: they are located in several nuclei of the brain stem, hypothalamus, and basal forebrain. The second consists of providing these management regions with moment-by-moment signals originating in all parts of the organism. These signals offer the managing regions — the control panels — a constantly updated view of the state of the organism.

Some of the signals are ferried directly by nerve pathways and signify the state of the viscera (e.g., heart, blood vessels, skin) or the muscles. Other signals come in the bloodstream and are conveyed by the concentration of a hormone or of glucose or of oxygen and carbon dioxide or by the pH level of the plasma. These signals are "read" by a number of neural sensing devices which react differently according to the set points of their "reading" scales. One analogy for this operation is that of the thermostat in relation to climate control: certain temperature readings trigger a response (heating or cooling until the desired set point is reached); certain values trigger no response. You can imagine some parts of the central nervous system, for instance, in the brain stem and hypothalamus, as a vast field of thermostat-like detectors, whose states of activity constitute a map. There are some perils in this analogy since the set points in a living organism can undergo changes across a lifetime and can be influenced in part by the context in which the sensing devices operate. After all, our thermostat-like detectors are made of living tissue, not of metal or silicone. For these reasons, Steven Rose has argued persuasively for the use of the word *homeodynamics* rather than *homeostasis*.[3] Nonetheless, the essence of the analogy is sound.

Why Are Body Representations Well Suited to Signify Stability?

The reason why representations of the body are well suited to signify stability comes from the remarkable invariance of the structures and operations of the body. Throughout development, adulthood, and even senescence, the *design* of the body remains largely unchanged. To be sure, bodies grow in size during development, but the fundamental systems and organs are the same throughout the life span and the operations that most components perform change little or not at all. This is generally true of bones, joints, and muscles, and especially true of the viscera and of the internal milieu. The range of possible states of the internal milieu and of the viscera is tightly limited. This limitation is built into the organism specifications since the range of states that is compatible with life is small. The permissible range is indeed so small and the need to respect its limits so absolute for survival that organisms spring forth equipped with an automatic regulation system to ensure that life-threatening deviations do not occur or can be rapidly corrected.

In short, not only is a considerable part of the body notable for its minimal variation—one might even say relative sameness—but living organisms naturally carry devices designed to insure limited variance, or if you will, maintain sameness. Those devices are planted genetically in any living being and do their basic job whether beings want it or not. Most beings do not "want" anything whatsoever, but in those that do, it makes no difference: the basic regulatory devices still operate in the same way.

So, if you are looking for a haven of stability in the universe of change that is the world of our brains, you might do worse than consider the regulating devices which maintain life in check, along with the integrated neural representations of the internal milieu, the viscera, and the musculoskeletal frame which portray the living state. Internal milieu, viscera, and musculoskeletal frame produce a continuous representation, dynamic but of narrow range, while the world around us changes dramatically, profoundly, and often unpredictably. Moment by moment, the brain has available a dynamic representation of an entity with a limited range of possible states—the body.[4]

ONE BODY ONE PERSON: THE ROOTS OF THE SINGULARITY OF SELF

You might wish to consider an amusing piece of evidence, at this point. For every person that you know, there is a body. You may never have given any thought to this simple relationship but there it is: one person, one body; one mind, one body—a first principle. You have never met a person without a body. Nor have you met a person with two bodies or with multiple bodies, not even Siamese twins. It just does not happen. You may have met, or heard about, bodies occasionally inhabited by more than one person, a pathological condition known as multiple personality disorder (it has a new name these days: dissociative identity disorder). Even then, however, the principle is not quite violated since, at each given time, only one among the multiple identities can use the body to think and behave, only one at a time gains enough control to *be* a person and express itself (better still, to express *its* self). The fact that multiple personalities are not considered normal reflects the general agreement that one body goes with one self.

One of the reasons we so admire good actors is that they can convince us that they are other persons, that they have other minds and other selves. But we know that they do not, we know that they are mere vessels for crafty make-believe, and we prize their work because what they do is neither natural nor easy.

Now, this is intriguing, is it not? Why should we not commonly find two or three persons in one body? What an economy of biological tissue. Or why should not persons of great intellectual capacity and imagination inhabit two or three bodies? What fun, what world of possibilities. Why should there not be bodiless persons in our midst, you know, ghosts, spirits, weightless and colorless creatures? Think of the space savings. But the simple fact is that such creatures do not exist now and nothing indicates that they ever did, and the sensible reason why not is that a mind, that which defines a person, requires a body, and that a body, a human body to be sure, naturally generates one mind. A mind is so closely shaped by the body and destined to serve it that only one mind could possibly arise in it. No body, never mind. For any body, never more than one mind.

Body-minded minds help save the body. When creatures like us appeared, which had bodies and conscious minds, they were, as Nietzsche would call them, "hybrids of plants and of ghosts," the combination of a bounded, well-circumscribed, easily identifiable living object with a seemingly unbounded, internal, and difficult-to-localize mental animation. He also called those creatures "discords," for they did possess a strange marriage of the clearly material with the apparently insubstantial. The marriage has puzzled everyone for millennia, and may now be, to some extent, a little easier to understand than before. Maybe.[5]

THE ORGANISM'S INVARIANCE AND THE IMPERMANENCE OF PERMANENCE

It is astonishing to discover that the seemingly rock-solid stabilities behind a single mind and a single self are themselves ephemeral and continuously reconstructed at the level of cells and molecules. This strange situation—an apparent rather than real paradox—has a simple explanation: although the building blocks for the construction of our organisms are regularly replaced, the architectural designs for the varied structures of our organisms are carefully maintained. There is a *Bauplan* for life and our bodies are a *Bauhaus*. Consider the following.

We are not merely perishable at the end of our lives. Most parts of us perish during our lifetime only to be substituted by other perishable parts. The cycles of death and birth repeat themselves many times in a life span—some of the cells in our bodies survive for as little as one week, most for not more than one year; the exceptions are the precious neurons in our brains, the muscle cells of the heart, and the cells of the lens. Most of the components that do not get substituted—such as the neurons—get changed by learning. (In fact, nothing being sacred, even some neurons may get substituted.) Life makes neurons behave differently by altering, for instance, the way they connect with others. No component remains the same for very long, and most of the cells and tissues that constitute our bodies today are not the same we owned when we entered college. What remains the same, in good part, is the construction plan for our organism structure and the set points for the operation of its parts. Call it the spirit of the form and the spirit of the function.[6]

When we discover what we are made of and how we are put together, we discover a ceaseless process of building up and tearing down, and we realize that life is at the mercy of that never-ending process. Like the sand castles on the beaches of our childhood, it can be washed away. It is astonishing that we have a sense of self at all, that we have—that most of us have, some of us have—some continuity of structure and function that constitutes identity, some stable traits of behavior we call a personality. Fabulous indeed, amazing for certain, that you are you and I am me.

But the problem goes beyond perishability and renewal. Just as death and life cycles reconstruct the organism and its parts according to a plan, the brain reconstructs the sense of self moment by moment. We do not have a self sculpted in stone and, like stone, resistant to the ravages of time. Our sense of self is a state of the organism, the result of certain components operating in a certain manner and interacting in a certain way, within certain parameters. It is another construction, a vulnerable pattern of integrated operations whose consequence is to generate the mental representation of a living individual being. The entire biological edifice, from cells, tissues, and organs to systems and images, is held alive by the constant execution of construction plans, always on the brink of partial or complete collapse should the process of rebuilding and renewal break down. The construction plans are all woven around the need to stay away from the brink.

THE ROOTS OF INDIVIDUAL PERSPECTIVE, OWNERSHIP, AND AGENCY

Whatever happens in your mind happens in time and in space relative to the instant in time your body is in and to the region of space occupied by your body. Things are in or out of you. Those that are out of you are stationary or moving. The stationary can be close or far or somewhere in between. Moving things may be looming toward you or moving away from you or traveling in some trajectory that misses you, but your body is the reference. Moreover, experiential perspective not only helps situate real objects but also helps situate ideas, be they concrete or abstract. Experiential perspective is a source of metaphor in organisms endowed with such rich cognitive capacities as abundant conventional memory, working memory, language, and the manipulative capacities we subsume by the term intelligence. For example, the notion of self is "close to my heart" but the idea of a homunculus is "far from my liking." Ownership and agency are, likewise, entirely related to a body at a particular instant and in a particular space. The things you own are close to your body, or should be, so that they remain yours, and this applies to things, lovers, and ideas. Agency, of course, requires a body acting in time and space and is meaningless without it.

Imagine yourself crossing a street, and now picture an unexpected car driving fast in your direction. The point of view relative to the car that is coming toward you is the point of view of your body, and it can be no other. A person watching this scene from a window on the third floor of the building behind you has a different point of view: that of his or her body. The car approaches, and the position of your head and neck is altered as you orient in its direction, while your eyes move conjugately to focus on the rapidly evolving patterns formed in your retinas. A world of adjustments is in full swing, from the vestibular system, which originates in the inner ear, has to do with balance, and serves to indicate body position in space, to the machinery of the colliculi, which guides eye and head and neck movement with the help of brain-stem nuclei, to the occipital and parietal cortices, which modulate the process from high up. But this is not all. Having a car zooming toward you does cause an emotion called fear, whether you want it or not, and does change many things in the state of your organism—the gut, the heart, and the skin respond quickly, among many others. Let me suggest that the signaling of all the changes I enumerated above are the means to implement individual organism perspective in your mind. Notice that I am not saying they are the means for you to *experience* organism perspective yet, which would be the same as *knowing* of it. Experience or knowledge of something, in a word, consciousness, comes later. Many of the changes that take place as the car approaches are happening to the multidimensional brain representation of the body proper that existed fleetingly in the instants immediately before the episode began unfolding; they are happening *to* the proto-self in your organism. The person watching the scene from the third-floor window has a different perspective but undergoes similar formal changes in his or her proto-self.

I would say that perspective is continually and irrevocably built by the processing of signals from a variety of sources. First, from a specific perceptual apparatus—in the example, the optical images being formed in the two retinas. Second, from the varied adjustments that are simultaneously carried out by the different muscular sectors of the body and by the vestibular system. In the example, retinal images change rapidly as a result of the approaching object, but for them to remain in focus, there must be adjustments in the muscles that control the lens and the pupil; the muscles that control the position of the eyeball; and the muscles that control the head, the neck, and the trunk.[7] Finally, there are signals deriving from emotional responses to a particular object, which would be quite marked in the case of the fast-approaching car and include changes in the smooth musculature of viscera, occurring at varied sites of the body. Note that, depending on the object, there may be different proportions of musculoskeletal and emotional accompaniment, but both are always present. The presence of all these signals — in this particular example, from retinal images, from muscular-postural adjustments, and from muscular-visceral-endocrine adjustments — describes both the object as it looms *toward the organism* and part of the reaction of the organism *toward the object* as the organism regulates itself to maintain a satisfactory processing of the object.

There is no such thing as a *pure* perception of an object within a sensory channel, for instance, vision. The concurrent changes I have just described are *not* an optional accompaniment. To perceive an object, visually or otherwise, the organism requires both specialized sensory signals *and* signals from the adjustment of the body, which are necessary for perception to occur.[8]

The statement that there is no such thing as a pure perception holds true even in circumstances in which you are precluded from moving — if you were given an injection of curare, for instance. After an injection of curare, none of your skeletal muscles moves because curare blocks the nicotinic receptors for the neurotransmitter acetylcholine. Yet the "visceral" muscles involved in emotion can move freely because curare does not affect their muscarinic receptors for acetylcholine.

The statement also holds true when you are simply thinking of an object rather than actually perceiving it in the world outside your organism. Here is the reason why: The records we hold of the objects and events that we once perceived include the motor adjustments we made to obtain the perception in the first place and also include the emotional reactions we had then. They are all coregistered in memory, albeit in separate systems. Consequently, even when we "merely" think about an object, we tend to reconstruct memories not just of a shape or color but also of the perceptual engagement the object required and of the accompanying emotional reactions, regardless of how slight. Whether you are immobile from curarization or quietly daydreaming in the darkness, the images you form in your mind *always* signal to the organism its own engagement with the business of making images and evoke some emotional reactions. You simply cannot escape the *affectation* of your organism, motor and emotional most of all, that is part and parcel of having a mind.

The perspective for a melody you hear or for an object you touch is, quite naturally, the perspective of your organism because it is drawn on the modifications that your organism undergoes during the events of hearing or touching. As for the sense of ownership of images and the sense of agency over those images, they, too, are a direct consequence of the machinations which create perspective. They are inherent in those machinations as foundational sensory evidence. Later, our creative and educated brains eventually clarify the evidence in the form of subsequent inferences, which also become known to us.

The organism perspective with which images are formed is essential for the preparation of acts involving the objects depicted in the images. The correct perspective relative to the oncoming car is important to design the movement with which you escape it, and the same applies to the perspective for a ball you are supposed to catch with your hand. The automatic sense of individual agency is born there and then. Later, you can draw inferences to the same effect. The fact that you had interacted with an object in order to create images of it makes the thought of *acting on* the object easier to conceive.

We should note that having all these changes occur is not enough for consciousness to occur. Consciousness occurs when we know, and we can only know when we map the relationship of object and organism. Only then can we possibly discover that all of the reactive changes described above are taking place in our own organisms and are caused by an object.

THE MAPPING OF BODY SIGNALS

Among the great barriers to the understanding of the ideas explored here are the incomplete and often confused notions that prevail about somatic signaling and the somatosensory system, which is supposed to convey the signals. The word *somatosensory,* as its etymological derivation appropriately implies, describes the sensing of the *soma,* which is Greek for "body." But often the notion that *soma* conjures up is narrower than it should be. Unfortunately, what most often comes to mind upon hearing the words *somatic* or *somatosensory* is the idea of touch or the idea of muscle and joint sensation. As it turns out, however, the somatosensory system relates to far more than that and is actually not one single system at all. It is a combination of several subsystems, each of which conveys signals to the brain about the state of very different aspects of the body. It is apparent that these different signaling systems surfaced at different points in evolution. They use different machinery in terms of the nerve fibers that carry the signals from the body to the central nervous system, and they are also different in the number, type, and position of the central nervous system relays onto which they map their signals. In fact, one aspect of somatosensory signaling does not use neurons at all but rather chemical substances available in the bloodstream. In spite of these distinctions, the varied aspects of somatosensory signaling work in parallel and in fine cooperation to produce, at multiple levels of the central nervous system, from the spinal cord and brain stem to the cerebral cortices, myriad maps of the multidimensional aspects of the body state at any given moment.

To give an idea of what the subsystems do and how they are organized, I will group the signaling into three fundamental divisions: the internal milieu and visceral division; the vestibular and musculoskeletal division; and the fine-touch division.

All three divisions can work both in close cooperation and in relative independence. When you touch an object whose texture gives you pleasure, signals from all three divisions have been brought to maps in the central nervous system which describe the ongoing interaction along its many dimensions, e.g., the movements with which you investigate the object; the properties which activate tactile sensors; and the humoral and visceral reactions which constitute the pleasurable response to the object. But the divisions may operate independently, e.g., the first with little help from the second, or the first and second with no help from the third. The important point to note is that the first division—the one concerned with the organism's interior—is permanently active, permanently signaling the state of the most internal aspects of the body proper to the brain. Under no normal condition is the brain ever excused from receiving continuous reports on the internal milieu and visceral states, and under most conditions, even when no active movement is being performed, the brain is also being informed of the state of its musculoskeletal apparatus. The brain is truly the body's captive audience as I noted.

The internal milieu and visceral division is in charge of sensing changes in the chemical environment of cells throughout the body. The term *interoceptive* describes those sensing operations generically. One aspect of these signals dispenses with nerve fibers and pathways altogether. Chemicals flowing in the bloodstream are sensed by nuclei of neurons in some regions of the brain stem, hypothalamus, and telencephalon. If the concentration of the chemical is within the permissible range, nothing happens. If the concentration is too high or too low, the neurons respond—they initiate a variety of actions aimed at achieving a correction of the imbalance. For instance, they can make you calm or make you jittery, they can make you feel hungry or wish to have sex, which is all fascinating, of course, but the point is that the signals create, moment by moment, multiple maps of the internal milieu, as many as the dimensions of our interior that can be measured with this peculiar method, and there are *many* such dimensions.

The brain's exposure to the chemicals that circulate in the bloodstream is remarkable. The brain is protected from the penetration of certain molecules by the so-called blood-brain barrier, a biological filter that envelops virtually all the blood vessels that carry nutrients to the brain tissue and is quite selective about what is or is not allowed to trespass from the blood into the brain tissue. A few brain regions, however, are devoid of blood-brain barrier and easily admit large molecules that, elsewhere in the brain, are kept from influencing the neural tissue directly. Molecules that cross the blood-brain barrier act on the brain directly, at sites like the hypothalamus; large molecules that cannot penetrate the blood-brain barrier get to act on the brain at special sites in which the barrier is missing, the so-called circumventricular organs. Examples of such sites are the area postrema (located in the brain stem) and the subfornical organs (located at cerebral hemisphere level). The chemically excited neurons in these areas pass their messages on to other neurons. The action of substances such as oxytocin, which is critical for a variety of behaviors, from sex and bonding to childbirth, depends on this arrangement. The brain's immersion in the chemical milieu is serious business.

The internal milieu and visceral division uses nerve pathways to carry the signals which we eventually perceive as pain, which can originate almost anywhere in the body, e.g., in viscera of the abdomen or in a joint or muscle. That division also carries neural signals related to aspects of the internal milieu so that the organism's chemical profile gets to be mapped not just via the bloodstream but also via neural pathways—for instance, pH levels and the concentration of oxygen and carbon dioxide are both dually mapped.

Finally, this division also signals the state of the smooth muscles, which are so abundant throughout the viscera and which are under autonomic control. The designation *autonomic* means that a particular process is controlled in its virtual entirety by devices independent of our will which are located in the brain stem, hypothalamus, and limbic nuclei, rather than in the cerebral cortex. There are smooth muscles everywhere, for example, in any blood vessel anywhere in the body. Those smooth muscles can contract or dilate to regulate blood circulation and its attendant functions. One result of such contraction or dilation of smooth muscle becomes well known to us when it increases or decreases systemic blood pressure or when it causes skin to blanch or to flush. Incidentally, the largest of all viscera in the body is the skin itself. I am not referring to the surface of the skin, which has a critical role in the sense of touch, but to the "thick of the skin," which is vital to the regulation of temperature. Extensive burns can kill you not because you lose tactile functions but because your homeostatic regulation is severely disturbed. This critical part of the skin's function derives from the ability to change the caliber of the many blood vessels that crisscross its thickness. "I've got you under my skin" unwittingly captures this important physiological idea and the lyric would have been even more accurate if Cole Porter had written, "I've got you in the thick of my skin," and it would be just as naughty. Predictably, the French have got it right on when they say "*Je t'ai dans la peau,*" which means, "I have you *in* my skin."

The signals I have been considering travel via a particular sector of the spinal cord (the lamina I and II of the posterior horn) and of the trigeminal nerve nucleus (the pars caudalis). I should add, however, that the convenient grouping of all these signals in one large division hides much in terms of channel subdivision. For instance, we know from the work of A. Craig that the neurons that carry signals related to nociception (pain) are different from those which mediate other aspects of body sense, although all of them draw on C-fibers and A-δ fibers.[9] On the other hand, we also know that many body-related signals are not only conveyed separately to high levels of the nervous system but also mixed and pooled together shortly after entering the central nervous system. This is what happens, for example, in the deeper zones of each spinal-cord segment.[10] Additional information for this division of the somatosensory system comes from viscera and is carried by visceral afferents to the spinal cord and by nerves such as the vagus nerve (which bypasses the cord altogether and aims directly at the brain stem).

The second division, the musculoskeletal one, conveys to the central nervous system the state of the muscles which join moving parts of the skeleton, that is, bones. When muscle fibers contract, the length of a muscle is reduced and the appropriately connected bones are pulled into motion. When muscle fibers relax, the opposite occurs. All the muscles that perform skeletal movement can be controlled by our will and are striated muscles (there is an exception to this rule and it has to do with the heart, whose muscle fibers are striated rather than smooth and yet are neither under volitional control nor in charge of moving any bony parts). The function of this division of the somatosensory system is generically known by the terms *proprioceptive* or *kinesthetic*. Just as is the case with the interoceptive signals from the internal milieu and viscera, proprioceptive/kinesthetic signals form many maps of the body aspects which they survey. These maps are placed at multiple levels of the central nervous system, all the way from the spinal cord to the cerebral cortex. The vestibular system, which maps the coordinates of the body in space, completes the somatosensory information under this division.

A third division of the somatosensory system conveys *fine touch*. Its signals describe the alterations which specialized sensors in the skin go through when we make contact with another object and investigate its texture, form, weight, temperature, and so on. While the internal milieu and visceral division is largely preoccupied with the description of internal states, the fine-touch division is mostly preoccupied with the description of external objects based on the signals generated in the body surface. The musculoskeletal division, somewhat in between, can be used both to express internal states as well as to help describe the outside world.

THE NEURAL SELF

The sense of self, in either core or autobiographical versions, is unlikely to have been the original variety of the phenomenon. I propose that the sense of self has a preconscious biological precedent, the *proto-self,* and that the earliest and simplest manifestations of self emerge when the mechanism which generates core consciousness operates on that nonconscious precursor.

The proto-self is a coherent collection of neural patterns which map, moment by moment, the state of the physical structure of the organism in its many dimensions. This ceaselessly maintained first-order collection of neural patterns occurs not in one brain place but in many, at a multiplicity of levels, from the brain stem to the cerebral cortex, in structures that are interconnected by neural pathways. These structures are intimately involved in the process of regulating the state of the organism. The operations of acting on the organism and of sensing the state of the organism are closely tied. The proto-self is not to be confused with the rich sense of self on which our current knowing is centered this very moment. *We are not conscious of the proto-self.* Language is not part of the structure of the proto-self. The proto-self has no powers of perception and holds no knowledge.[11]

Nor is the proto-self to be confused with the rigid homunculus of old neurology. The proto-self does not occur in one place only, and it emerges dynamically and continuously out of multifarious interacting signals that span varied orders of the nervous system. Besides, the proto-self is not an interpreter of anything. It is a reference point at each point in which it is.

This hypothesis should be considered in the perspective of an important qualification regarding the relation between brain regions and functions, such as proto-self. Such functions are not "located" in one brain region or set of regions, but are, rather, a product of the interaction of neural and chemical signals among a set of regions. This is true of the nonconscious proto-self in relation to the set of regions I outline below, and it is also true of functions such as core self or autobiographical self, to be discussed later. Phrenological thinking must be resisted at all costs.

The structures required to implement the proto-self are listed below, along with those which are not required to implement it. Drawing on the two lists, it is possible to test the hypothesis in a variety of ways. The most direct way consists of formulating predictions regarding the effects of damage to some of the key structures presented in both lists. Some lesions ought to disrupt the proto-self and consequently disrupt consciousness, more or less severely, while others ought to leave consciousness unscathed. A preliminary assessment of the validity of those predictions is possible on the basis of current evidence from neuropathology and neurophysiology but further prospective studies are needed to firm up any conclusions.

Brain Structures Required to Implement the Proto-Self

1. Several *brain-stem nuclei* which regulate body states and map body signals. Along the chains of signaling that begin in the body and terminate in the highest and most distal structures of the

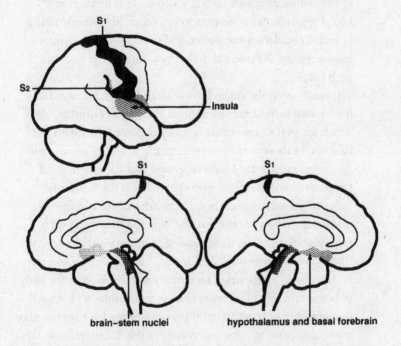

Figure 5.1. Location of some proto-self structures. Note that the region known as the insula is buried inside the sylvian fissure and not visible on the cortical surface.

brain, this region is the first in which an aggregate of nuclei signal the overall current body state, as mediated by the spinal cord pathways, the trigeminal nerve, the vagus complex, and the area postrema. Included in this region are classical reticular nuclei as well as monoamine and acetylcholine nuclei.[12]

2. The *hypothalamus,* which is located near the structures named in 1 and closely interconnected with them, and the *basal forebrain,* which is located in the vicinity of the hypothalamus, is interconnected with both hypothalamus and brain stem, and constitutes an extension of those lower structures into the forebrain. The hypothalamus contributes to the current representation of the body by maintaining a current register of the state of the internal milieu along several dimensions, e.g., level of circulating nutrients such as glucose, concentration of varied ions, relative concentration of water, pH, concentration of varied circulating hormones, and so on. The hypothalamus helps regulate the internal milieu by acting on the basis of such maps.

3. The *insular cortex,* the *cortices known as S2,* and the *medial parietal cortices* located behind the splenium of the corpus callosum, all of which are part of the somatosensory cortices. In humans the function of these cortices is asymmetric. Based on my own observations in patients, I have suggested that the ensemble of these cortices in the right hemisphere holds the most integrated representation of the current internal state of the organism at the level of the cerebral hemispheres, along with representations of the invariant design of the musculoskeletal frame. In a recently published article, Jaak Panksepp also links body and self, by means of an innate representation of the body in brain stem. His idea comes close to my notion of proto-self, in several respects, although his view of how such a representation contributes to consciousness is entirely different from mine.[13]

Brain Structures Which Are Not Required to Implement the Proto-Self

The structures listed below are not required to implement the proto-self. This non-exhaustive list covers most of the central nervous system. It includes all the early sensory cortices for external sensory modalities—which means that it includes visual and auditory cortices as well as the sectors of somatosensory cortices concerned with fine touch; all the temporal and most of the frontal higher-order cortices (higher-order cortices are those that are not exclusively dedicated to one sensory modality but rather to supramodal integration of signals related to early sensory cortices), and the hippocampal formation and its interconnected cortices, e.g., entorhinal cortex (area 28) and the perirhinal cortices (area 35). The specific roster is as follows:

1. Several early sensory cortices, namely those of areas 17, 18, 19, which are dedicated to vision; 41/42, 22, dedicated to hearing; area 37, which is partly dedicated to vision but is also a higher-order cortex (see 2, below), and the part of S1 concerned with fine touch. These cortices are involved in the making of modality-specific sensory patterns, which support the mental images of diverse sensory modalities available in our mind. They play a role in consciousness, both core and extended, inasmuch as the object to be known is assembled from these regions, but they play no role in the proto-self.

2. All the inferotemporal cortices, namely areas 20, 21, part of 37, 36, and 38. These cortices are the basis for the dispositional (implicit) memories that can be reconstructed in recall in the form of explicit sensory patterns and mental images. These cortices support many of the autobiographical records on the basis of which the autobiographical self can be assembled and extended consciousness realized.

3. The hippocampus, a vital structure in the "on-line" mapping of multiple, concurrent stimuli. The hippocampus receives signals related to activity in all sensory cortices, which arrive

indirectly at the end of several projection chains with multiple synapses, and reciprocates signals via backward projections along the same chains. It is essential to create new memories of facts but not new memories of perceptuomotor skills. It appears to hold memories within itself temporarily but not permanently. Most importantly, it appears to contribute to the establishment of memories elsewhere, in circuitry connected to it.

4. The hippocampal-related cortices, namely areas 28 and 35. These cortices may hold dispositional memories of even higher complexity than those in 2, above.

5. The prefrontal cortices. A vast array of higher-order cortices. Some of them hold high-complexity dispositions for personal memories involving unique temporal and spatial contexts; for memories of the relation between certain categories of events or entities and somatic states; and for memories for abstract concepts. Some of these cortices participate in high-level working memory for spatial, temporal, and linguistic functions. Because of their role in working memory, prefrontal cortices are critical for high levels of extended consciousness. Because of their role in autobiographical memory, they are relevant to autobiographical self and extended consciousness.

6. The cerebellum. One of the most transparent but also elusive sectors of the brain. It is obviously involved in the construction of fine movement—you cannot shoot straight without it, never mind sing, play an instrument, or play tennis. Yet it is also involved in affective and cognitive processes, and I suspect all the more so during development. It may be involved in the processes of emotion and of mental search, e.g., searching for a specific word or nonverbal item in memory. The lack of severe dysfunction following its ablation or inactivation suggests that the role it plays in cognition is subtle. But recent studies suggest this could be an artifact of inadequate observation, made

all the more likely by the cerebellum's blatant anatomical and functional redundancy.

SOMETHING-TO-BE-KNOWN

We have seen how a specific set of neural structures can support the first-order representation of current body states that I call the proto-self, and in so doing, provide the roots for the self, the "something-to-which-knowing-is-attributed." It is time to say something about the roots for the other key player in the process: the "something-to-be-known."

The background for our understanding of how the brain represents the something-to-be-known is extensive. We have a considerable, though incomplete, understanding of how sensory representations in the main sensory modalities (e.g., vision, hearing, touch) are related to signals arising in peripheral sensory organs, such as the eye or the inner ear, and how those signals are relayed to the respective primary sensory regions of the cerebral cortex by means of subcortical nuclei such as those in the thalamus. Beyond the primary sensory cortices we understand a little about how explicit mental representations— those which have a manifest structure—are related to varied neural maps and about how some memory for those representations can be recorded in implicit manner. We know, for instance, that varied aspects of an object—for instance, its form, its color and motion, or the sounds it produces—are handled in a relatively segregated way by cortical regions located downstream from the respective primary visual or auditory cortices. We suspect that some kind of neural integrative process helps generate, within the overall region related to each modality—the so-called early sensory cortices—the composite of neural activities which support the integrated image we experience.[14] However, we do not know all the intermediate steps between neural patterns and mental patterns. We do know that the same overall region supports image making for both perception (which we construct from the actual scene external to the brain, from the outside in) and for re-call (which we reconstruct in the mind internally, inside out, as it were). We have reasons to believe that the integration of sensory rep-resentations across modalities—say, vision *and* auditory, or vision *and* touch—may well depend on timing mechanisms that coordinate ac-tivity across large regions of the brain and probably will not need yet another single integrative space per se—a single Cartesian theater. And we know for certain that basic sensory integration does not re-quire higher-order cortices in anterior temporal and prefrontal cor-tices.[15] (See the appendix, section 3, for a more extensive discussion of these issues.)

Let us now consider first the situation of an actual something-to-be-known, an actual object. Such an object is implemented in early sensory cortices, those collections of cortices in which signals from the varied sensory channels, such as vision, hearing, and touch, are processed along the many dimensions of an object, such as color, shape, motion, auditory frequencies, and so on.

The presence of such signals from an actual object provokes in the organism the sort of response I discussed earlier in this chapter, namely, a collection of motor adjustments required to continue gathering signals about the object as well as emotional responses to several aspects of the object. In other words, the implementation of the something-to-be-known is inevitably accompanied by a com-plex effect on the proto-self, that is, an effect on the very neural basis of the something-to-which-knowing-is-attributed. Let me repeat that this is enough for *being* but not enough for *knowing*, that is, not enough to be conscious. Consciousness, as we shall see, only arises when the object, the organism, and their relation, can be re-represented.

Now let us turn to the case of an object that is not actually present but has, rather, been committed to memory. According to my frame-work, the memory of that object has been stored in dispositional form. Dispositions are records which are dormant and implicit rather than active and explicit, as images are. Those dispositional memories of an object that was once actually perceived include not only records of the sensory aspects of the object, such as the color, shape, or sound, but also records of the motor adjustments that necessarily accompanied the gathering of the sensory signals; moreover the memories also contain records of the obligate emotional reaction to the object. As a consequence, when we recall an object, when we allow dispositions to make their implicit information explicit, we retrieve not just sensory data but also accompanying motor and emotional data. When we recall an object, we recall not just sensory characteristics of an actual object but the past reactions of the organism to that object.

The significance of the distinction between actual object and memorized object will become clear in the next chapter. I will preview that significance by saying that this distinction permits memorized objects to engender core consciousness in the same way that actually perceived objects do. This is why we can be conscious of what we remember as much as we are conscious of what we actually see, hear, or touch now. Were it not for this magnificent arrangement, we could never have developed an autobiographical self.

A Note on the Disorders of the Something-to-Be-Known

The disorders of the something-to-be-known fall into two broad categories: perceptual disorders and agnosias. In perceptual disorders, a lack of signals from a sensory modality such as vision or hearing or the somatosensory division of touch prevents the sensory representation of an object from being formed—acquired blindness or deafness are examples. Under those circumstances, an object X, which was to be represented by a particular sensory channel, can no longer be represented, fails to engage the organism in the usual manner, and does not modify the proto-self. The result is that no core consciousness ensues.

Now for the second category, the agnosias. *Agnosia* is an obscure but well-formed word that denotes an inability to conjure up from memory the sort of knowledge that is pertinent to a given object as the object is being perceived. The percept is stripped of its meaning, as an old and lapidary definition stated so well. The exemplary form of agnosia is the condition known as *associative agnosia,* to use technical neurological terms. Associative agnosia occurs with respect to the main sensory modalities, e.g., there are cases of visual agnosia, auditory agnosia, and tactile agnosia. Because of their exquisite specificity, these are some of the most intriguing cases encountered in neurology. As you will discover in the illustration below, a perfectly sane and intelligent human being can be deprived of the ability to recognize familiar persons by sight but not by sound (or vice versa).

It Must Be Me because I'm Here

That is what Emily said cautiously as she contemplated the face in the mirror before her. It had to be her; she had placed herself in front of the mirror, of her own free will, so it had to be her; who else could it be? And yet she could not recognize her face in the looking glass; it was a woman's face, all right, but whose? She did not think it was hers and she could not confirm it was hers since she could not bring her face back into her mind's eye. The face she was looking at did not conjure up anything specific in her mind. She could believe it was hers because of the circumstances: She had been brought by me into this room and asked to walk to the mirror and see who was there. The situation told her unequivocally that it could not be anyone else and she accepted my statement that, of course, it was her.

Yet, when I pressed "play" on the tape deck and let her hear an audiotape of her own voice, she immediately recognized it as hers. She had no difficulty recognizing her unique voice even if she could no longer recognize her unique face. This same disparity applied to everyone else's faces and voices. She could not recognize her husband's face, her children's faces, or the faces of other relatives, friends, and acquaintances. However, she could easily recognize their characteristic voices.

Emily was not unlike David in the sense that "nothing came to mind" when certain specific items were shown to her. But she was vastly different in the sense that her problem pertained exclusively to the visual world. Nothing came to mind *only* when she was shown the visual aspect of a unique stimulus with whom or with which she was perfectly familiar—a person's face, a particular house, a particular vehicle. The nonvisual aspects of the same stimulus—say, sound or touch—brought to mind everything they were supposed to bring.[16]

Emily did better with the less than unique. Remarkably, she could easily tell that a face whose identity she could no longer access expressed an emotion. The same was true of the age and gender of the person who owned a certain face.[17] Her problem was confined to unique items in the visual medium.

How does Emily fare on my core consciousness checklist? The answer is, perfectly. I do not need to tell you that she is awake and attentive in every way. Her attention focuses easily and is sustained for all sorts of tasks. Her emotions and the feelings she reports are entirely normal, too. Her behavior is purposeful and appropriate for all contexts, immediate as well as long term, limited only by her visual difficulties. In fact, even in spite of those difficulties, she can do remarkable intellectual feats. She sits for hours observing people's gaits and tries to guess who they are, often successfully; she can hold perfect conversations with guests at the receiving line of her parties, provided her husband whispers the name of the visually unknown person; and she can find her visually unrecognizable car in the supermarket parking lot by checking systematically all the license plates.

I do want to call your attention to something quite revealing, however. Not only is she conscious of what she knows perfectly well, but she is also conscious of what she does not know. She generates core consciousness for every stimulus that comes her way regardless of the amount of knowledge she can conjure up about the stimulus. Emily, as well as the many other patients like her that I have studied over the years, is perfectly conscious of the things she does *not* know and she examines those things, in reference to her knowing self, in the same way she examines the things she does know. Consider the following experiment we customized for Emily.

We had noted, purely by chance, as we used a long sequence of photographs to test her recognition of varied people, that upon looking at the photo of an unknown woman who had one upper tooth slightly darker than the rest, Emily ventured that she was looking at her daughter.

"Why do you think it is your daughter?" I remember asking her.

"Because I know Julie has a dark upper tooth," she said. "I bet it is her."

It wasn't Julie, of course, but the mistake was revealing of the strategy our intelligent Emily now had to rely on. Unable to recognize identity from global features and from sets of local features of the face, Emily seized upon any simple feature that could remind her of anything potentially related to any person she might be reasonably asked to recognize. The dark tooth evoked her daughter and on that basis she made an informed guess that it was indeed her daughter.

To check on the validity of this interpretation, we designed a simple experiment. We modified a few photos of smiling men and women so that they would show a slightly darker upper incisor and interspersed them randomly in a stack of many other photos. Whenever Emily came to a modified photo of any young woman—never the men or older women—she proclaimed it to be her daughter. She had a keen awareness for the whole and for the parts of the photos she was shown, or she would have had no possibility of reasoning as intelligently as she did, item after item, and would have had no chance of spotting the target stimuli. In the very least, Emily and those like her demonstrate that one does not require specific knowledge of an item at a unique level in order to have core consciousness of the item.

When a patient with face agnosia fails to recognize the familiar face in front of her and affirms that she has never seen that person, that she has no recollection of anything related to that person, the pertinent knowledge is not being deployed for conscious survey, but core consciousness remains intact. In fact, once you confront the patient with the fact that the face before her is that of a close friend, the patient is not only conscious in general but conscious also of her failure,

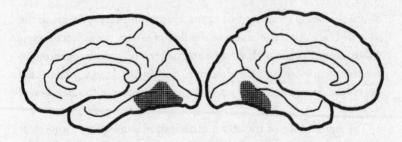

Figure 5.2. The lesions that caused prosopagnosia in patient Emily were located at the junction of the occipital and temporal lobes of both hemispheres. This is the typical location of lesions in patients with associative prosopagnosia.

conscious of her inability to conjure up any knowledge useful to recognize the close friend. Her problem is not one of consciousness but of memory. The specific something-to-be-known is missing—she cannot represent the knowledge of who it is she is looking at, she cannot be conscious of something now present. But core consciousness is present as generated by other layers of something-to-be-known—for instance, the face as face, as opposed to the face of a unique person. It is precisely because normal core consciousness is present that the recognition void comes to be acknowledged.

Emily's problem was caused by bilateral damage in the early visual cortices, specifically in the visual association cortices located at the transition of occipital and temporal lobes in the ventral aspect of the brain. Brodmann's areas 19 and 37, in a region known as the fusiform gyrus, bore the brunt of the damage.

On the basis of our early neuroimaging correlations regarding face agnosia, almost two decades ago, we suggested that these cortices were normally involved in the processing of faces and of other visually ambiguous stimuli that made similar demands on the brain.[18] Current functional neuroimaging experiments support this idea: normal individuals consistently activate the region damaged in Emily's brain when they are aware of processing a face.[19] It is important to note that activation of this area in a functional neuroimaging experiment should not be interpreted as meaning that "consciousness for faces" occurs in the so-called face area. The image of the face of which the subject is conscious cannot occur without a neural pattern becoming organized in the face area, *but the remainder of the process that generates the sense of knowing that face and that drives attention to the pattern is occurring elsewhere, in other components of the system.*

The significance of the above qualification is nowhere more clear than when we consider the following fact: when an unconscious patient in persistent vegetative state was shown familiar faces, the so-called "face area" (at the occipito-temporal junction, within the fusiform gyrus) lit up in a functional imaging scan, much as it does in normal and sentient persons.[20] The moral of this story is simple: the power to make neural patterns for the something-to-be-known is preserved even when consciousness is no longer being made.

BILATERAL DAMAGE TO auditory cortices yields the same results as damage to visual cortices as far as core consciousness goes. In the same way that Emily does not conjure up specific knowledge pertinent to unique items, such as the previously familiar person or object, patients with damage within selected regions of the auditory sector of the cerebral cortex lose the ability to conjure up specific knowledge pertinent to, say, a previously familiar melody or the previously familiar voice of a unique person. The patient known in my laboratory as patient X. illustrates the situation. He is a highly accomplished and successful opera singer who, as a result of a stroke, lost the ability to recognize the singing voices of the colleagues with whom he had performed around the world. As for his own singing voice, he could no longer recognize it, either. He also lost the ability to identify familiar melodies including those of arias he had sung hundreds of times in his long career. Just as was the case with Emily, he had no problem outside the auditory realm and, just as was the case with Emily, he properly generated core consciousness for the stimuli that he was no longer able to know in the proper sense of the term. He scrutinized each unrecognized piece with keen awareness, searching within every tone, within its color and mode of production, for a possible clue to the identity of the singer producing it. The only voice he was ever able to recognize unfailingly was that of Maria Callas, perhaps one more bit of evidence that Callas was indeed a breed apart.

Both Emily and X. have damage within the association cortices, respectively visual and auditory association cortices. It is apparent, then, from the study of numerous cases like theirs, that extensive damage in those sensory cortices does not compromise core consciousness. When it comes to extensive damage of early sensory cortices, only damage to the somatosensory regions causes a disruption of consciousness, for the reasons adduced earlier: the somatosensory regions are part of the basis of the proto-self, and their damage can easily alter the basic mechanisms of core consciousness.

NOW THAT WE know how the brain can put together the neural patterns that represent an object, and the neural patterns that represent an individual organism, we are ready to consider the mechanisms that the brain may use to represent the relationship between the object and the organism — the causal action of the object on the organism and the resulting possession of the object by the organism.

Chapter Six

The Making of
Core Consciousness

The Birth of Consciousness

How do we ever begin to be conscious? Specifically, how do we ever have a sense of self in the act of knowing? We begin with a first trick. The trick consists of constructing an account of what happens within the organism when the organism interacts with an object, be it actually perceived or recalled, be it within body boundaries (e.g., pain) or outside of them (e.g., a landscape). This account is a simple narrative without words. It does have characters (the organism, the object). It unfolds in time. And it has a beginning, a middle, and an end. The beginning corresponds to the initial state of the organism. The middle is the arrival of the object. The end is made up of reactions that result in a modified state of the organism.

We become conscious, then, when our organisms internally construct and internally exhibit a specific kind of wordless knowledge—that our organism has been changed by an object—and when such knowledge occurs along with the salient internal exhibit of an object. The simplest form in which this knowledge emerges is the feeling of knowing, and the enigma before us is summed up in the following question: By what sleight of hand is such knowledge gathered, and why does the knowledge first arise in the form of a feeling?

The specific answer I deduced is presented in the following hypothesis: *core consciousness occurs when the brain's representation devices generate an imaged, nonverbal account of how the organism's own state is affected by the organism's processing of an object, and when this process enhances the image of the causative object, thus placing it saliently in a spatial and temporal context.* The hypothesis outlines two component mechanisms: the generation of the imaged nonverbal account of the object-organism relationship—which is the source of the sense of self in the act of knowing—and the enhancement of the images of an object. As far as the sense-of-self component is concerned, the hypothesis is grounded on the following premises:

1. Consciousness depends on the internal construction and exhibition of new knowledge concerning an interaction between that organism and an object.
2. The organism, as a unit, is mapped in the organism's brain, within structures that regulate the organism's life and signal its internal states continuously; the object is also mapped within the brain, in the sensory and motor structures activated by the interaction of the organism with the object; both organism and object are mapped as neural patterns, in first-order maps; all of these neural patterns can become images.
3. The sensorimotor maps pertaining to the object cause changes in the maps pertaining to the organism.
4. The changes described in 3 can be re-represented in yet other maps (second-order maps) which thus represent the relationship of object and organism.
5. The neural patterns transiently formed in second-order maps can become mental images, no less so than the neural patterns in first-order maps.

6. Because of the body-related nature of both organism maps and second-order maps, the mental images that describe the relationship are feelings.

I note, again, that the focus of our inquiry here is not the matter of how neural patterns in any map become mental patterns or images—that is the *first* problem of consciousness as outlined in chapter 1. We are focusing on the *second* problem of consciousness, the problem of self.

As far as the brain is concerned, the organism in the hypothesis is represented by the proto-self. The key aspects of the organism addressed in the account are those I indicated as provided in the proto-self: the state of the internal milieu, viscera, vestibular system, and musculoskeletal frame. The account describes the relationship between the changing proto-self and the sensorimotor maps of the object that causes those changes. In short: As the brain forms images of an object—such as a face, a melody, a toothache, the memory of an event—and as the images of the object *affect* the state of the organism, yet another level of brain structure creates a swift nonverbal account of the events that are taking place in the varied brain regions activated as a consequence of the object-organism interaction. The mapping of the object-related consequences occurs in first-order neural maps representing proto-self and object; the account of the *causal relationship* between object and organism can only be captured in second-order neural maps. Looking back, with the license of metaphor, one might say that the swift, second-order nonverbal account narrates a story: *that of the organism caught in the act of representing its own changing state as it goes about representing something else.* But the astonishing fact is that the knowable entity of the catcher has just been created in the narrative of the catching process.

This plot is incessantly repeated for every object the brain represents, and it does not matter whether the object is present and interacting with the organism or is being brought back from past memory. It also makes no difference what the object really is. In healthy individuals, as long as the brain is awake, the machines of image making and consciousness are "on," and we are not manipulating our mental state by doing something like meditation, it is not possible to run out of "actual" objects or "thought" objects, and it is thus not possible to run out of the abundant commodity called core consciousness. There are just too many objects, actual or recalled, and often there is more than one object at about the same time. The same imaged plot is supplied in abundance to the flowing process we call thought.[1]

The wordless narrative I propose is based on neural patterns which become images, images being the same fundamental currency in which the description of the consciousness-causing object is also carried out. Most importantly, the images that constitute this narrative are incorporated in the stream of thoughts. The images in the consciousness narrative flow like shadows along with the images of the object for which they are providing an unwitting, unsolicited comment. To come back to the metaphor of movie-in-the-brain, they are *within* the movie. There is no external spectator.[2]

Now let me conclude my presentation of how I think core consciousness arises, by addressing the second component in the hypothesis. The process which generates the first component—the imaged nonverbal account of the relationship between object and organism—has two clear consequences. One consequence, already presented, is the subtle image of knowing, the feeling essence of our sense of self; the other is the enhancement of the image of the causative object, which dominates core consciousness. Attention is driven to focus on an object and the result is saliency of the images of that object in mind. The object is *set out* from less-fortunate objects—selected as a particular *occasion* in both the Jamesian and Whiteheadian senses. It becomes *fact*, following the preceding events which lead to its becoming, and it is part of a relationship with the organism to which all this is happening.

You Are the Music while the Music Lasts: The Transient Core Self

You know that you are conscious, you feel that you are in the act of knowing, because the subtle imaged account that is now flowing in the stream of your organism's thoughts exhibits the knowledge that your proto-self has been changed by an object that has just become salient in the mind. You know you exist because the narrative exhibits you as protagonist in the act of knowing. You rise above the sea level of knowing, transiently but incessantly, as a *felt* core self, renewed again and again, thanks to anything that comes from outside the brain into its sensory machinery or anything that comes from the brain's memory stores toward sensory, motor, or autonomic recall. You know it is *you* seeing because the story depicts a character— you—doing the seeing. The first basis for the conscious *you* is a feeling which arises in the re-representation of the *nonconscious proto-self in the process of being modified* within an account which establishes the cause of the modification. The first trick behind consciousness is the creation of this account, and its first result is the feeling of knowing.

Knowing springs to life in the story, it inheres in the newly con- structed neural pattern that constitutes the nonverbal account. You hardly notice the storytelling because the images that dominate the mental display are those of the things of which you are now con- scious—the objects you see or hear—rather than those that swiftly constitute the feeling of you in the act of knowing. Sometimes all you notice is the whisper of a subsequent verbal translation of a related in- ference of the account: Yes, it is me seeing or hearing or touching. But, faint as it may be, half guessed as the hint often is, when the sto- rytelling is suspended by neurological disease, your consciousness is suspended as well and the difference is monumental.[3]

T. S. Eliot might as well have been thinking of the process I just de- scribed when he wrote, in the *Four Quartets,* of "music heard so deeply that it is not heard at all," and when he said "you are the music while the music lasts." He was at least thinking of the fleeting moment in which a deep knowledge can emerge—a union, or incarnation, as he called it.

Beyond the Transient Core Self: The Autobiographical Self

Something does last after the music is gone, however; some residue does remain after many ephemeral emergences of core self. In com- plex organisms such as ours, equipped with vast memory capacities, the fleeting moments of knowledge in which we discover our existence are facts that can be committed to memory, be properly categorized, and be related to other memories that pertain both to the past and to the anticipated future. The consequence of that complex learning operation is the development of autobiographical memory, an aggregate of dispositional records of who we have been physically and of who we have usually been behaviorally, along with records of who we plan to be in the future. We can enlarge this aggregate memory and refashion it as we go through a lifetime. When certain personal records are made explicit in reconstructed images, as needed, in smaller or greater quantities, they become the *autobiographical self.* The real marvel, as I see it, is that autobiographical memory is architecturally connected, neurally and cognitively speaking, to the nonconscious proto-self and to the emergent and conscious core self of each lived instant. This connection forms a bridge between the ongoing process of core consciousness, condemned to sisyphal transiency, and a progressively larger array of established, rock-solid memories pertaining to unique historical facts and consistent characteristics of an individual. In other words, the body-based, dynamic-range stability of the nonconscious proto-self, which is reconstructed live at each instant, and the conscious core self, which emerges from it in the second-order nonverbal account when an object modifies it, are enriched by the accompanying display of memorized and invariant facts—for instance, where you were born, and to whom; critical events in your autobiography; what you like and dislike; your name; and so on. Although the basis for the autobiographical self is stable and invariant, its scope changes continuously as a result of experience. The display of autobiographical self is thus more open to refashioning than the core self, which is reproduced time and again in essentially the same form across a lifetime.

Unlike the core self, which inheres as a protagonist of the primordial account, and unlike the proto-self, which is a current representation of the state of the organism, the autobiographical self is based on a concept in the true cognitive and neurobiological sense of the term.

Table 6.1. Kinds of Self

AUTOBIOGRAPHICAL SELF: The autobiographical self is based on
autobiographical memory which is constituted by implicit memories of
multiple instances of individual experience of the past and of the anticipated
future. The invariant aspects of an individual's biography form the basis for
autobiographical memory. Autobiographical memory grows continuously
with life experience but can be partly remodeled to reflect new experiences.
Sets of memories which describe identity and person can be reactivated as a
neural pattern and made explicit as images whenever needed. Each reactivated
memory operates as a "something-to-be-known" and generates its own pulse
of core consciousness. The result is the autobiographical self of which we are
conscious.

CORE SELF: The core self inheres in the second-order nonverbal account that
occurs whenever an object modifies the proto-self. The core self can be triggered
by any object. The mechanism of production of core self undergoes minimal
changes across a lifetime. We are conscious of the core self.

CONSCIOUSNESS
_____

PROTO-SELF: The proto-self is an interconnected and temporarily coherent
collection of neural patterns which represent the state of the organism,
moment by moment, at multiple levels of the brain. We are *not* conscious of
the proto-self.

The concept exists in the form of dispositional, implicit memories
contained in certain interconnected brain networks, and many of
these implicit memories can be made explicit at any time, simultaneously.[4] Their activation in image form constitutes a backdrop to each
moment of a healthy mental life, usually unattended, often just
hinted and half guessed, just like the core self and like knowing, and
yet there, ready to be made more central if the need arises to confirm
that we are who we are. That is the material we use when we describe
our personality or the individual characteristics of another person's
mode of being. More about this in the next chapter when we discuss

Table 6.2. Distinguishing Core Self from Autobiographical Self

CORE SELF	AUTOBIOGRAPHICAL SELF
The transient protagonist of consciousness, generated for any object that provokes the core-consciousness mechanism. Because of the permanent availability of provoking objects, it is continuously generated and thus appears continuous in time.	Based on permanent but dispositional records of core-self experiences. Those records can be activated as neural patterns and turned into explicit images. The records are partially modifiable with further experience.
The mechanism of core self requires the presence of proto-self. The biological essence of the core self is the representation in a second-order map of the proto-self being modified.	The autobiographical self requires the presence of a core self to begin its gradual development.
	The autobiographical self also requires the mechanism of core consciousness so that activation of its memories can generate core consciousness.

extended consciousness and the mechanisms behind identity and personhood.

In a developmental perspective, I expect that in the early stages of our being, there is little more than reiterated states of core self. As experience accrues, however, autobiographical memory grows and the autobiographical self can be deployed. The milestones that have been identified in child development are possibly a result of the uneven expansion of autobiographical memory and the uneven deployment of the autobiographical self.[5]

Regardless of how well autobiographical memory grows and how robust the autobiographical self becomes, it should be clear that they require a continued supply of core consciousness for them to be of any consequence to their owner organism. The contents of the autobiographical self can only be known when there is a fresh construction of core self and knowing for each of those contents to be known. A patient in the throes of an epileptic automatism has not destroyed her autobiographical memory and yet cannot access its contents. When the seizure ends and core consciousness returns, the bridge is reestablished and the autobiographical self can be called up as needed. In other words, although the contents of the autobiographical self pertain to the individual in a most unique way, they depend on the gift of core consciousness to come alive just as any other something-to-be-known. A bit unfair, perhaps, but that is how it must be.

ASSEMBLING CORE CONSCIOUSNESS

I see core consciousness as created in pulses, each pulse triggered by each object that we interact with or that we recall. Let's say that a consciousness pulse begins at the instant just before a new object triggers the process of changing the proto-self and terminates when a new object begins triggering its own set of changes. The proto-self modified by the first object then becomes the *inaugural* proto-self for the new object. A new pulse of core consciousness begins.

The continuity of consciousness is based on the steady generation of consciousness pulses which correspond to the endless processing of myriad objects, whose interaction, actual or recalled constantly, modifies the proto-self. The continuity of consciousness comes from the abundant flow of nonverbal narratives of core consciousness.

It is probable that more than one narrative is created simultaneously. This is because more than one object can be engaged at about the same time, although not many can be engaged simultaneously, and more than one object can thus induce a modification in the state of the proto-self. When we talk about a "stream of consciousness," a metaphor that suggests a single track and a single sequence of thoughts, the part of the stream that carries consciousness is likely to arise not in just one object but in several. Moreover, it is also probable that each object interaction generates more than one narrative, since several brain levels can be involved. Again, such a situation seems beneficial because it would produce an overabundance of core consciousness and ensure the continuity of the state of "knowing." I shall say some more on the issue of multiple generators of core consciousness in the pages ahead.

THE NEED FOR A SECOND-ORDER NEURAL PATTERN

Telling the story of the changes caused on the inaugural proto-self by the organism's interaction with any object requires its own process and its own neural base. In the simplest terms, I would say that beyond the many neural structures in which the causative object and the proto-self changes are separately represented, there is at least one other structure which *re-represents* both proto-self and object in their temporal relationship and can thus represent what is actually happening to the organism: *proto-self at the inaugural instant; object coming into sensory representation; changing of inaugural proto-self into proto-self modified by object.* I suspect, however, that there are several structures in the human brain with the ability to generate a second-order neural pattern which re-represents first-order occurrences. The second-order neural pattern which subtends the nonverbal imaged account of the organism-object relationship is probably based on intricate cross-signaling among several "second-order" structures. The likelihood is low that one brain region holds *the* supreme second-order neural pattern.

The main characteristics of the second-order structures whose interaction generates the second-order map are as follows: A second-order structure must (1) be able to receive signals via axon pathways signals from sites involved in representing the proto-self *and* from sites that can potentially represent an object; (2) be able to generate a neural pattern that "describes," in a temporally ordered manner, the events occurring in the first-order maps; (3) be able, directly or indirectly, to introduce the image resulting from the neural pattern in the overall flow of images we call thought; and (4) be able, directly or indirectly, to signal back to the structures processing the object so that the object image can be enhanced.

A.

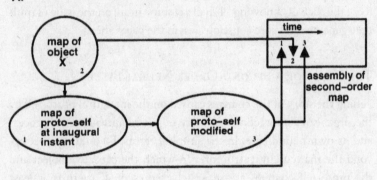

B.

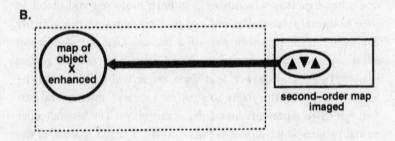

Figure 6.1. A. Components of the second-order neural pattern assembled in temporal sequence in second-order structure. B. Second-order map image arises and map of object becomes enhanced.

A sketch of this general idea is presented in figure 6.1. A second-order structure receives a succession of signals related to an unfolding event that occurs at different brain sites—the forming of the image of object X; the state of the proto-self as the image of X begins to be formed; the changes in the proto-self caused by processing X. This succession of re-representations constitutes a neural pattern that becomes, directly or indirectly, the basis for an image—the image of a relationship between object X and the proto-self changed by X. Let me stress, again, that this is a simplification of the idea. In all likelihood, because there are several second-order structures, the neural pattern and the image of the relationship will result from the cross-signaling among those second-order structures. Also note that, as we saw earlier, the process of core consciousness is not confined to generating this imaged account. The presence of the account pattern in a second-order neural pattern has important consequences: it influences the neural maps of the object by modulating their activity and thus enhances the saliency of those patterns for a brief period.

Where Is the Second-Order Neural Pattern?

It is important to consider the possible anatomical sources for the second-order pattern. My best guess is that the second-order neural pattern arises transiently out of interactions among a select few regions. It is not to be found within a single brain region—some sort of phrenologically conceived consciousness center—but neither is it everywhere or anywhere. The fact that the second-order neural pattern is implemented in more than one site may sound surprising at first, but it should not. I believe it conforms to a general brain rule rather than to an exception. Consider, for instance, what happens with movement. Imagine yourself in a room when a friend enters and wishes to borrow a book. You get up and walk over, picking up the book as you do, and begin talking; your friend says something amusing; you begin laughing. You are producing movements with your whole body, as you rise and begin your trajectory, and as a certain posture is being adopted for that purpose; your legs are moving and so is your right arm; so are parts of your speech apparatus; so are the muscles in your face, rib cage, and diaphragm as you laugh. As in the analogy of behavior as orchestral performance, there are half a dozen *separate* motor generators, each doing its part, some under voluntary control (the ones that help you pick up the book), others not (the ones that control body posture or laughter). All of them, however, are beautifully coordinated in time and space so that your movements are smoothly performed and appear generated by a single source and by a single will. We have few clues as to how and where this astonishing smoothing and blending occurs. No doubt it all occurs with the help of a slew of brain stem, cerebellar, and basal ganglia circuits, interacting by cross-signaling. Precisely how is not clear, of course.

Now transfer the above conditions to my conception of core consciousness. Here, too, I am suggesting that there are multiple consciousness generators, at several brain levels, and yet the process appears smooth, concerned with one knower and one object. It is reasonable to assume that under normal circumstances several second-order maps relative to different aspects of the processing of an object would be created in parallel, roughly within the same time interval. Core consciousness for that object would result from a composite of second-order maps, an integrated neural pattern which would give rise to the imaged account I proposed earlier and also lead to the enhancement of the object. I do not know how the fusing, blending, and smoothing are achieved, but it is important to note that the mystery is not particular to consciousness; it pertains to other functions such as motion. Perhaps when we solve the latter, we also solve the former.

There are several brain structures capable of receiving converging signals from varied sources and thus seemingly capable of second-order mapping. In the context of the hypothesis, the second-order structures I have in mind must achieve a specific conjoining of signals from "whole-organism maps" and "object maps." Respecting such demands relative to the source of the signals to be conjoined eliminates several candidates, e.g., higher-order cortices in the parietal and temporal regions, the hippocampus, and the cerebellum, whose roles fall under first-order mapping. Moreover, the second-order structures required by the hypothesis must be capable of exerting an influence on first-order maps so that enhancement and coherence of object images can occur. Once this other demand is also taken into account, the real contenders for second-order structure are the superior colliculi (the twin hill-like structures in the back part of the midbrain known as the tectum); the entire region of the cingulate cortex; the thalamus; and some prefrontal cortices. I suspect that all of these contenders play a

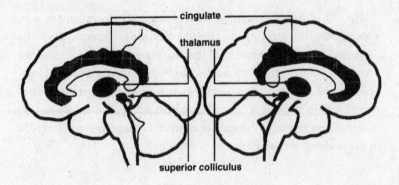

Figure 6.2. Location of the main second-order map structures, mentioned in the hypothesis.

role in consciousness; that none of them acts alone; and that the scope of their contributions is varied. For example, I doubt the superior colliculi are especially important in human consciousness, and I suspect the prefrontal cortices probably participate only in extended consciousness. Figure 6.2 gives a rough idea of where these structures are.

The notion of interaction among such structures is critical to the hypothesis. For example, as regards core consciousness, I believe that both superior colliculi and cingulate cortices independently assemble a second-order map. Yet, the second-order neural pattern I envision in my hypothesis as the basis for our feeling of knowing is supraregional. It would result from the ensemble playing of the superior colliculi and the cingulate under the coordination of the thalamus, and it is sensible to assume that the cingulate and thalamic components would have the lion's share in the ensemble.

The subsequent influence of second-order neural patterns on the enhancement of the object image is achieved by several means, including thalamocortical modulation and the activation of acetylcholine and monoamine nuclei in the basal forebrain and brain stem, all of which subsequently affect cortical processing. It is interesting to note that the second-order structures I propose would indeed have the means to exert such influences.

The list of neuroanatomical devices required to implement consciousness is thus growing but remains mercifully circumscribed. The list includes the select number of structures needed to implement the proto-self (some brain-stem nuclei, the hypothalamus and basal forebrain, some somatosensory cortices) as well as the structures enumerated here as possible second-order mapping sites. In chapter 8 I consider how plausible the involvement of all these structures may be in the making of consciousness.

THE IMAGES OF KNOWING

The first use for the imaged account of the organism-object relationship is to inform the organism of what it is doing, or put in different words, to answer a question that was never posed by the organism: What is happening? What is the relation between images of things and this body? The feeling of knowing is the beginning of the answer. I have already outlined the consequences of acquiring such unsolicited knowledge: it is the beginning of the freedom to *comprehend* a situation, the beginning of the eventual chance to plan responses that differ from the Duchampian "ready-mades" provided by nature.

As I suggested, however, there is an immediate secondary use for the process that leads to the imaged account. When the properly equipped brain of a wakeful organism generates core consciousness, the first result is *more* wakefulness—note that some wakefulness was available already and was necessary to start the ball rolling. The second result is *more-focused* attention to the causative object—again, some attention was available already. Both results are achieved by means of enhancing the first-order maps which represent the object.

To some degree, the message implied in the conscious state is: "Focused attention must be paid to X." Consciousness results in *enhanced* wakefulness and *focused* attention, both of which improve image processing for certain contents and can thus help optimize immediate and planned responses. The organism's engagement with an object intensifies its ability to process that object sensorily and also increases the opportunity to be engaged by other objects—the organism gets ready for more encounters and for more-detailed interactions. The overall result is greater alertness, sharper focus, higher quality of image processing.

Beyond providing a feeling of knowing and an enhancement of the object, the images of knowing, assisted by memory and reasoning, form the basis for simple nonverbal inferences which strengthen the process of core consciousness. These inferences reveal, for instance, the close linkage between the regulation of life and the processing of images which is implicit in the sense of individual perspective. Ownership is hidden, as it were, within the sense of perspective, ready to be made clear when the following inference can be made: if these images have the perspective of this body I now feel, then these images are in my body—they are mine. As for the sense of action, it is contained in the fact that certain images are tightly associated with certain options for motor response. Therein our sense of agency—these images are mine and I can act on the object that caused them.

CONSCIOUSNESS FROM PERCEIVED OBJECTS AND RECALLED PAST PERCEPTIONS

When objects appear in mind not because they are immediately present in our surroundings but because we recall them from memory, their images also cause core consciousness. The reason for this has to do with the fact that we store in memory not just aspects of an object's physical structure—the potential to reconstruct its form, or color, or sound, or typical motion, or smell, or what have you—but also aspects of our organism's motor involvement in the process of apprehending such relevant aspects: our emotional reactions to an object; our broader physical and mental state at the time of apprehending the object. As a consequence, recall of an object and deployment of its image in mind is accompanied by the reconstruction of at least some of the images which represent those pertinent aspects. Reconstructing that collection of organism accommodations for the object you recall generates a situation similar to the one that occurs when you perceive an external object directly.[6]

The net result is that as you think about an object, reconstructing part of the accommodations required to perceive it in the past as well as the emotive responses to it in the past is enough to change the proto-self in much the same manner that I have described for when an external object confronts you directly. The immediate source of the object of which you become conscious is different, in actual perception or recall, but the consciousness of apprehending something is the same, whether perceived or recalled. This is why curarized patients, who are unable to produce actual somatomotor postural adjustments in order to perceive an object, are still mentally aware of objects brought to their stationary sensory devices. In all likelihood, even the plans for future perceptuo-motor accommodations are effective modifiers of the proto-self and thus originators of second-order accounts. If both the actions themselves as well as the plans for actions can be the source of second-order maps, then core consciousness can arise even earlier since plans for movement necessarily occur before movements, just as the responses that eventually cause emotions occur before those emotions are enacted.

Because our brain has the possibility of representing, in somatosensory maps, both plans of action and actions themselves, and because such plans can be made available to second-order maps, the brain would have available a double mechanism for constructing the primordial narratives of consciousness.

THE NONVERBAL NATURE OF CORE CONSCIOUSNESS

Let me make clear what I mean by making a narrative or telling a story. The terms are so connected to language that I must ask you again not to think of them in terms of words. I do not mean narrative or story in the sense of putting together words or signs in phrases and sentences. I do mean telling a narrative or story in the sense of creating a nonlanguaged map of logically related events. Better to think of film (although the film medium does not give the perfect idea, either) or of mime—Jean-Louis Barrault miming the story of the watch theft in *Les Enfants du Paradis*. A line from a poem by John Ashbery captures the idea: "This is the tune but there are no words, the words are only speculation (from the Latin *speculum*)."[7]

In the case of humans the second-order nonverbal narrative of consciousness can be converted into language immediately. One might call it the third-order. In addition to the story that signifies the act of knowing and attributes it to the newly minted core self, the human brain also generates an automatic verbal version of the story. I have no way of stopping that verbal translation, neither do you. Whatever plays in the nonverbal tracks of our minds is rapidly translated in words and sentences. That is in the nature of the human, languaged creature. This uninhibitable verbal translation, the fact that knowing and core self *also* become verbally present in our minds by the time we usually focus on them, is probably the source of the notion that consciousness might be explainable by language alone. It has been thought that consciousness occurred when, and only when, language commented on the mental situation for us. As indicated earlier, the view of consciousness required by this notion suggests that only humans with substantial mastery of the language instrument would have conscious states. Nonlanguaged animals and human babies would be just out of luck, forever unconscious.

The language explanation of consciousness is improbable and we need to go behind the mask of language to find a more plausible alternative. Curiously, the very nature of language argues against it having a primary role in consciousness. Words and sentences denote entities, actions, events, and relationships. Words and sentences translate concepts, and concepts consist of the nonlanguage idea of what things, actions, events, and relationships are. Of necessity, concepts precede words and sentences in both the evolution of the species and the daily experience of each and every one of us. The words and sentences of healthy and sane humans do not come out of nowhere, cannot be the de novo translation of nothing before them. So when my mind says "I" or "me," it is translating, easily and effortlessly, the nonlanguage concept of the organism that is mine, of the self that is mine. If a perpetually activated construct of core self were not in place, the mind could not possibly translate it as "I," or as "me," or as whatever literary paraphrase it might apply, in whatever language it might know. The core self must be in place for its translation into a suitable word to occur.

One could argue, in fact, that the consistent content of the *verbal* narrative of consciousness—regardless of the vagaries of its form—permits one to deduce the presence of the equally consistent *nonverbal, imaged* narrative that I am proposing as the foundation of consciousness.

The narrative of the state of the proto-self being changed by the interaction with an object must first occur in its nonlanguage form if it is ever to be translated by suitable words. In the sentence "I see a car coming," the word *see* stands for a particular act of perceptual possession perpetrated by my organism and involving my self. And the word *see* is there, properly moored to the word *I,* to translate the wordless play unfolding in my mind.

Now let me say that my views could be questioned along the following lines. What if the wordless play of core consciousness, the nonverbal narrative of knowing, occurs below the level of consciousness and only the verbal translation provides evidence that it occurs at all? Core consciousness would emerge only at the time of verbal translation and not before, during the nonverbal phase of the storytelling. The view I find less plausible would be brought back but with a small twist: the mechanisms I outlined to describe the actors and events in the act of knowing would remain, but the possibility that the nonverbal narrative alone would give us access to knowing would be denied.

This alternative view would be interesting, but I am not ready to endorse it. The main reason not to do so comes from the need to rely on language and on its powers in order to have consciousness. To begin with, although verbal translations cannot be inhibited, they are often not attended, and they are performed under considerable literary license—the creative mind translates mental events in a large variety of ways rather than in a stereotypical manner. Moreover, the creative "languaged" mind is prone to indulge in fiction. Perhaps the most important revelation in human split-brain research is precisely this: that the left cerebral hemisphere of humans is prone to fabricating verbal narratives that do not necessarily accord with the truth.[8]

I find it unlikely that consciousness would depend on the vagaries of verbal translation and on the unpredictable level of focused attention paid to it. If consciousness depends on verbal translations for its existence, chances are one would have varying kinds of consciousness, some truthful, some not; varied levels and intensity of consciousness, some effective, some not; and, worst of all, lapses of consciousness. Yet this is not what happens in healthy and sane humans. The primordial story of self and knowing is told with consistency. Your degree of *focused attention* to an object does vary, but your level of general consciousness does not drop below threshold when you are distracted from an object and focus on another—you do not become stuporous and it does not look like you are having a seizure; you are just conscious of other things rather than conscious of nothing. The threshold of consciousness is met when you wake up, and after that, consciousness stays on until it is turned off. When you run out of words and sentences, you do not fall asleep: you just listen and watch.

I believe the imaged, nonverbal narrative of core consciousness is swift, that its unexamined details have eluded us for a long time, that the narrative is barely explicit, so half hinted that its expression is almost like the emanation of a belief. But some aspects of the narrative filter into our minds to create the beginning of the knowing mind and the beginnings of the self. Those aspects, captured in the feeling of self and knowing, are the first above the sea level of consciousness and precede the corresponding verbal translation.

Requiring consciousness to depend on the presence of language leaves no room for core consciousness as I have outlined here. Consciousness, according to the language-dependency hypothesis, follows language mastery and thus cannot occur in organisms that lack that mastery. When Julian Jaynes presents his engaging thesis about the evolution of consciousness, he is referring to consciousness post-language, not to core consciousness as I described it. When thinkers as diverse as Daniel Dennett, Humberto Maturana, and Francisco Varela speak about consciousness, they usually refer to consciousness as a postlanguage phenomenon. They are speaking, as I see it, about the higher reaches of extended consciousness as it occurs now, at this stage in biological evolution.[9] I have no problem with their proposals, but I wish to make clear that, in my proposal, extended consciousness rides on top of the foundational core consciousness which we and other species have long had and continue to have.

THE NATURALNESS OF WORDLESS STORYTELLING

Wordless storytelling is natural. The imagetic representation of sequences of brain events, which occurs in brains simpler than ours, is the stuff of which stories are made. A natural preverbal occurrence of storytelling may well be the reason why we ended up creating drama and eventually books, and why a good part of humanity is currently hooked on movie theaters and television screens. Movies are the closest external representation of the prevailing storytelling that goes on in our minds. What goes on within each shot, the different framing of a subject that the movement of the camera can accomplish, what goes on in the transition of shots achieved by editing, and what goes on in the narrative constructed by a particular juxtaposition of shots is comparable in some respects to what is going on in the mind, thanks to the machinery in charge of making visual and auditory images, and to devices such as the many levels of attention and working memory.

Be that as it may, the marvel is to think that the very first brains that constructed the story of consciousness were answering questions that no living being had yet posed: Who is making these images that have just been happening? Who owns these images? "*Who's there?*," as in the stirring first line of *Hamlet*, a play that so powerfully epitomizes the bewilderment of humans regarding the origins of their condition.[10] The answers had to come first, by which I mean that the organism had to construct first the kind of knowledge that looks like answers. The organism had to be able to produce that primordial knowledge, unsolicited, so that a process of knowing could be founded.

The entire construction of knowledge, from simple to complex, from nonverbal imagetic to verbal literary, depends on the ability to map what happens over time, *inside* our organism, *around* our organism, *to* and *with* our organism, one thing followed by another thing, causing another thing, endlessly.

Telling stories, in the sense of registering what happens in the form of brain maps, is probably a brain obsession and probably begins relatively early both in terms of evolution and in terms of the complexity of the neural structures required to create narratives. Telling stories precedes language, since it is, in fact, a condition for language, and it is based not just in the cerebral cortex but elsewhere in the brain and in the right hemisphere as well as the left.[11]

Philosophers often puzzle about the problem of so-called "intentionality," the intriguing fact that mental contents are "about" things outside the mind. I believe that the mind's pervasive "aboutness" is rooted in the brain's storytelling attitude. The brain inherently represents the structures and states of the organism, and in the course of regulating the organism as it is mandated to do, the brain naturally weaves wordless stories about what happens to an organism immersed in an environment.

ONE LAST WORD ON THE HOMUNCULUS

A comment on the infamous homunculus solution for the problem of self, and on why it failed, is in order at this point. The disqualified homunculus solution consisted of postulating that a part of the brain, "the knower part," possessed the knowledge needed to interpret the images formed in that brain. The images were presented to the knower, and the knower knew what to do with them. In this solution, the knower was a spatially defined container, the so-called homunculus. The term suggested the picture that many people actually formed of its physical structure: a little man scaled down to the confines of brain size. Some even imagined the homunculus to look like the familiar drawing that appears in the textbook diagrams of the motor and somatosensory regions of the cerebral cortex, the one with the tongue sticking out and the feet upside down.

The problem with the homunculus solution was that the all-knowing little person would do the knowing for each of us but would then face the difficulty with which we began in the first place. Who would do *its* knowing? Well, another little person, of course, only smaller. In turn, the second little person would need a third little person inside to be its knower. The chain would be endless and this postponing of the difficulty, known as infinite regress, effectively disqualified the homunculus solution. This disqualification was a good thing, of course, inasmuch as it emphasized the inadequacy of a traditional brain "center" account for something as complex as knowing. But it had a chilling effect on the development of alternate solutions. It created a fear of the homunculus, worse than the fear of flying, which eventually became the fear of specifying a knowing self, cognitively and neuroanatomically. In short order, the act of knowing and self went from being inside a little brain person to being nowhere.

The failure of the homunculus idea to provide a solution for how we know cast doubt on the very notion of self. This was unfortunate. One should, indeed, be skeptical of a homunculus-like knower, endowed with full knowledge and located in a single and circumscribed part of the brain. It makes no sense physiologically. All the available evidence suggests that nothing like it exists. The failure of the homunculus-style knower, however, does not suggest that the notion of self should or could be discarded along with that of the homunculus. Whether we like the notion or not, something like the sense of self does exist in the normal human mind as we go about knowing things. Whether we like it or not, the human mind is constantly being split, like a house divided, between the part that stands for the known and the part that stands for the knower.

The story contained in the images of core consciousness is not told by some clever homunculus. Nor is the story really told by *you* as a self because the core *you* is only born as the story is told, *within the story itself.* You exist as a mental being when primordial stories are being told, and only then; as long as primordial stories are being told, and only then. You are the music while the music lasts.

Brains equipped with the appropriate devices beyond the well-known sensory and motor devices can form images of the organism caught in the act of forming images of other things and reacting to those images. Those extra devices permit the act of knowing in an organism previously equipped with the ability to represent a stable proto-self and to represent a great many things that can happen within its body proper and to it. There is no homunculus involved. There is also no regress of any kind, infinite or otherwise. In the homunculus-knower version of consciousness, a special knowing agency is asked to please explain what is going on; the neural-mental-knower homunculus must know more than the brain/mind it serves. But, of course, then comes the next knower homunculus who must know more than the previous one, and on we go *ad infinitum* and *absurdum.* In my proposal there is no need to interrogate any agency, any knower. Moment by moment, the answer is being presented to the organism, as represented by the proto-self, placed before it in the form of a nonverbal narrative which can be subsequently translated in a language. The explanation is presented prior to any request for it.

The proto-self is a reference rather than a storehouse of knowledge or an intelligent perceiver. It participates in the process of knowing, waiting patiently for a most generous brain to explain what is happening by answering questions that were never posed: Who *does*? Who *knows*? When the answer first arrives, the sense of self emerges, and to us now, creatures endowed with rich knowledge and an autobiographical self, millions of years after the first instances of primordial storytelling ever occurred, it does appear as if the question was posed, and that the self is a knower who knows.

No questions asked then. There is no need to interrogate the core self about the situation and the core self does not interpret anything. Knowing is generously offered free of charge.

TAKING STOCK

I have been proposing that core consciousness depends on a ceaselessly generated image of the act of knowing, first expressed as a feeling of knowing relative to the mental images of the object to be known; and I also proposed that the feeling of knowing results in, and is accompanied by, an enhancement of the images of the object.

Turning to the possible biology behind core consciousness, I proposed a set of neural structures and operations which may support the emergence of the sense of self and of knowing. The proposal, presented in the form of a hypothesis, was designed to meet the requisites outlined for the biological role of consciousness and for the description of its mental appearance as well as to conform to known facts of neuroanatomy and neurophysiology. The hypothesis states that core consciousness occurs when the brain forms an imaged, non-verbal, second-order account of how the organism is causally affected by the processing of an object. The imaged account is based on second-order neural patterns generated from structures capable of receiving signals from other maps which represent both the organism (the proto-self) and the object.[12]

The assembly of the second-order neural pattern describing the object-organism relationship modulates the neural patterns which describe the object and leads to the enhancement of the image of the object. The comprehensive sense of self in the act of knowing an object emerges from the contents of the imaged account, *and* from the

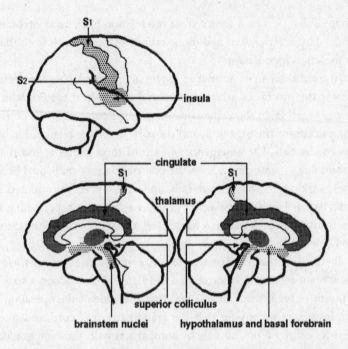

Figure 6.3. The main proto-self and second-order map structures combined. Note that most of these structures are located near the brain's midline.

enhancement of the object, presumably in the form of a large-scale pattern that combines both components in a coherent manner.

The neuroanatomical structures required by the hypothesis encompass those that support the proto-self; those needed to process the object; and those needed to generate the imaged account of the relationship and to produce its consequences.

The neuroanatomy underlying the processes behind proto-self and object (presented in chapter 5) includes brain-stem nuclei, the hypothalamus, and somatosensory cortices. The neuroanatomy underlying the imaged account of the relationship and the enhancement of object image (presented earlier in this chapter) includes the cingulate cortices, the thalamus, and the superior colliculi. The subsequent image enhancement is achieved via modulation from basal forebrain/ brain-stem acetylcholine and monoamine nuclei as well as from thal- amocortical modulation.

In conclusion, in its normal and optimal operation, core conscious- ness is the process of achieving a neural and mental pattern which brings together, in about the same instant, the pattern for the object, the pattern for the organism, and the pattern for the relationship be- tween the two. The emergence of each of those patterns and their conjoining in time depends on the contributions of individual brain sites working in close cooperation, and in the proposal outlined in this chapter I address one aspect of the overall process, pertaining to the construction of patterns for the relationship between organism and object.

By the time the grand pattern of core consciousness emerges, a few local brain regions have succeeded in recruiting into action a sizable amount of brain tissue. If you find that the scale of the operation is impressive, now consider that the grand pattern of core conscious- ness is nothing if not humble by comparison with the even grander pattern of extended consciousness, to which I turn in the next chap- ter. Just as William James would have wished, nearly the whole brain is engaged in the conscious state.

Chapter Seven

Extended Consciousness

EXTENDED CONSCIOUSNESS

If core consciousness is the indispensable foundation of consciousness, extended consciousness is its glory. When we think of the greatness of consciousness we have extended consciousness in mind. When we slip and say that consciousness is a distinctively human quality, we are thinking of extended consciousness at its highest reaches, not of core consciousness, and we should be forgiven for the arrogance: extended consciousness is indeed a prodigious function, and, at its peak, it is uniquely human.

Extended consciousness goes beyond the here and now of core consciousness, both backward and forward. The here and now is still there, but it is flanked by the past, as much past as you may need to illuminate the now effectively, and, just as importantly, it is flanked by the anticipated future. The scope of extended consciousness, at its zenith, may span the entire life of an individual, from the cradle to the future, and it can place the world beside it. On any given day, if only you let it fly, extended consciousness can make you a character in an epic novel, and, if only you use it well, it can open wide the doors to creation.

Extended consciousness is everything core consciousness is, only bigger and better, and it does nothing but grow across evolution and across a lifetime of experience in each individual. If core consciousness allows you to know for a transient moment that it is you seeing a bird in flight or that it is you having a sensation of pain, extended consciousness places these same experiences in a broader canvas and over a longer period of time. Extended consciousness still hinges on the same core "you," but that "you" is now connected to the lived past and anticipated future that are part of your autobiographical record. Rather than just accessing the fact that you have pain, you can also survey the facts concerning where the pain is (the elbow), what caused it (tennis), when you last had it before (three years ago, or was it four?), who has also had it recently (Aunt Maggie), the doctor she went to (Dr. May, or was it Dr. Nichols?), the fact that you will not be able to play with Jack tomorrow. The range of knowledge that extended consciousness now allows you to access encompasses a large panorama. The self from which that large landscape is viewed is a robust concept in the true sense of the word. It is an autobiographical self.

The autobiographical self hinges on the consistent reactivation and display of selected sets of autobiographical memories. In core consciousness, the sense of self arises in the subtle, fleeting feeling of knowing, constructed anew in each pulse. Instead, in extended consciousness, the sense of self arises in the consistent, reiterated display of some of our own personal memories, the *objects of our personal past*, those that can easily substantiate our identity, moment by moment, and our personhood.

The secret of extended consciousness is revealed in this arrangement: autobiographical memories are *objects*, and the brain treats them as such, allows each of them to relate to the organism in the manner described for core consciousness, and thus allows each of them to generate a pulse of core consciousness, a sense of self know-ing. In other words, extended consciousness is the precious conse-quence of two enabling contributions: First, the ability to learn and thus retain records of myriad experiences, previously known by the power of core consciousness. Second, the ability to reactivate those records in such a way that, as objects, they, too, can generate "a sense of self knowing," and thus be known.

As one moves, biologically speaking, from the simple level of core consciousness, with its generic sense of self, to the complex levels of extended consciousness, the prime physiological novelty is memory for facts. As for the prime trick, it consists of more of the same: multiple generations of simple "sense of self knowing" applied *both* to the something-to-be-known and to an eternally revived and com-plex something-to-which-the-knowledge-is-attributed—the autobio-graphical self. The final enabling factor is working memory, the ability to hold active, over a substantial amount of time, the many "objects" of the moment: the object being known and the objects whose display constitutes our autobiographical self. The time scale is no longer the fraction of a second that characterizes core conscious-ness. We are now in the scale of seconds and minutes, the time scale at which most of our personal lives are transacted and which can easily extend to hours and years.

In short, extended consciousness emerges from two tricks. The first trick requires the gradual buildup of memories of many instances of a special class of objects: the "objects" of the organism's biography, of our own life experience, as they unfolded in our past, illuminated by core consciousness. Once autobiographical memories are formed, they can be called up whenever any object is being processed. Each of those autobiographical memories is then treated by the brain as an object, each becoming an inducer of core consciousness, along with the particular nonself object that is being processed. While relying on the same fundamental mechanism of core consciousness—the cre-ation of mapped accounts of ongoing relationships between organism and objects—extended consciousness applies the mechanism not just to a single nonself object X, but to a consistent set of previously memorized objects pertaining to the organism's history, whose relentless recall is consistently illuminated by core consciousness and constitutes the autobiographical self.

The second trick consists of holding active, simultaneously and for a substantial amount of time, the many images whose collection defines the autobiographical self and the images which define the object. The reiterated components of the autobiographical self and the object are bathed in the feeling of knowing that arises in core consciousness.

Extended consciousness is, then, the capacity to be aware of a large compass of entities and events, i.e., the ability to generate a sense of individual perspective, ownership, and agency, over a larger compass of knowledge than that surveyed in core consciousness. The sense of autobiographical self to which this larger compass of knowledge is attributed includes unique biographical information.

Autobiographical selves occur only in organisms endowed with a substantial memory capacity and reasoning ability, but do not require language. Developmental psychologists such as Jerome Kagan have suggested that humans develop a "self" by the time they are eighteen months old, and perhaps even earlier. I believe the self to which they refer is the autobiographical self.[1] I also believe apes such as bonobo chimpanzees have an autobiographical self, and I am willing to venture that some dogs of my acquaintance also do. They possess an autobiographical self but not quite a person. You and I possess both, of course, thanks to an even more ample endowment of memory, reasoning ability, and that critical gift called language. Over evolutionary time as well as individual time, our autobiographical selves have permitted us to know about progressively more complex aspects of the organism's physical and social environment and the organism's place and potential range of action in a complicated universe.

Extended consciousness is not the same as intelligence. Extended consciousness has to do with making the organism aware of the largest possible compass of knowledge, while intelligence pertains to the ability to manipulate knowledge so successfully that novel re-

Table 7.1. Kinds of Self

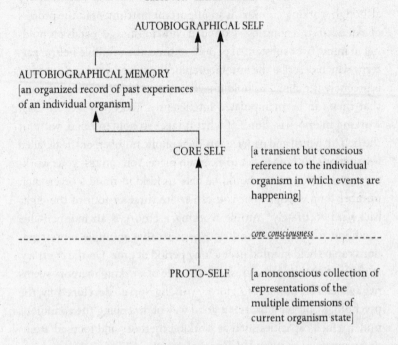

The arrow between the nonconscious proto-self and the conscious core self represents the transformation that occurs as a result of the mechanism of core consciousness. The arrow toward autobiographical memory denotes the memorization of repeated instances of core-self experiences. The two arrows toward autobiographical self signify its dual dependency on both continuous pulses of core consciousness and continuous reactivations of autobiographical memories.

---

sponses can be planned and delivered. Extended consciousness has to do with exhibiting knowledge and with displaying it clearly and efficiently so that intelligent processing can take place. Extended consciousness is a prerequisite of intelligence—how could one behave intelligently over vast domains of knowledge, if one could not survey such knowledge in extended consciousness?

Extended consciousness is also not the same as working memory although working memory is an important instrument in the process of extended consciousness. Extended consciousness depends on holding in mind, over substantial periods of time, the multiple neural patterns which describe the autobiographical self; and working memory is precisely the ability to hold images in mind for a long enough time that they can be manipulated intelligently. To get an idea of what working memory is, think of what it takes to hold in mind, without the help of pencil and paper, a ten-digit phone number, or the detailed instructions for how to get to a certain place. You can test your working memory, too: you should be able to hold in mind a seven-digit number long enough that you can recite three or four of the digits backward accurately.[2] Ample working memory is an indispensable condition for extended consciousness, so that multiple representations can be held in mind over a long period of time. On the contrary, at the level of core consciousness, the role of working memory seems negligible. The notion of "global working space" developed by the psychologist Bernard Baars is a good way of describing the conditions under which capacities such as working memory and focused attention contribute to extended consciousness.[3]

Core consciousness is part of the standard equipment of complex organisms such as we are; it is put in place by the genome with a little help from the early environment. Perhaps culture can modify it to some extent but probably not by much. Extended consciousness is also laid out by the genome, but culture can significantly influence its development in each individual.

Assessing Extended Consciousness

Extended consciousness is based on core consciousness not just for its development over time but moment by moment. The study of neurological patients shows that when core consciousness is removed, out goes extended consciousness. As we have seen, patients with absence seizures, epileptic automatisms, akinetic mutism, and persistent vegetative state have neither core consciousness nor extended consciousness. The converse is not true: as we will see in the pages ahead,

impairments of extended consciousness are compatible with preserved core consciousness.

Extended consciousness is a bigger subject than core consciousness,
and yet it is easier to address scientifically. We understand fairly well
what it consists of cognitively and we also understand the corresponding behavioral features. An organism in possession of extended
consciousness gives evidence of attention over a large domain of information which is present not just in the external environment but
also internally, in the environment of its mind. For example, as a possessor of extended consciousness, you are probably paying attention
to a number of different mental contents simultaneously: the printed
text; the ideas it evokes; questions it raises; perhaps music or a specific
noise somewhere in the house; and you yourself as knower. Not all
those contents are equally salient, equally sharply defined, but they
are all onstage, and at one time or another, over many seconds or
even minutes, one or a few come to the limelight.

An organism with extended consciousness gives evidence of planning of complex behaviors, not just on the moment but over larger
intervals of time—many hours and days, weeks and months. An observer can infer that such complex and appropriate behaviors were
planned by taking into account the history of the individual and the
current context. In other words, what a person does must make sense
not just in immediate terms but in terms of larger-scale contexts.

The work of Hans Kummer in baboons and of Marc Hauser in
chimps suggests that what I am describing as extended consciousness
is present in nonhuman species. Kummer's painstaking fieldwork
and Hauser's ingenious laboratory experiments reveal behaviors that
would require the cognitive operations described above. An example
is the elaborate and time-consuming decision-making behavior of a
troop of baboons concerned with choosing the place where they
should drink on a given day. Numerous factors impinge on the decision—for instance, estimated presence of water at the drinking site,
risk of encountering predators, distance, and so on. The evidence
suggests that those factors are heeded and connected to the homeostatic needs of the individuals.[4]

Extended consciousness is necessary for the internal deployment of a substantial amount of recalled knowledge in different sensory systems and modes, and for the subsequent abilities to manipulate that knowledge in problem solving or to report on it. The normal performance of all these abilities testifies to the presence of extended knowledge. The assessment of extended consciousness can be achieved by assessing recognition, recall, working memory, emotion and feeling, and reasoning and decision making over large intervals of time in an individual whose core consciousness is intact.

In a neurologically normal state, we are never completely deprived of extended consciousness. Yet it is not difficult to imagine what a possessor of *only* core consciousness probably experiences. Just consider what it may be like inside the mind of a one-year-old infant. I suspect objects come to the mind's stage, are attributed to a core self, and exit as quickly as they enter. Each object is known by a simple self and clear on its own, but there is no large-scale relation among objects in space or time and no sensible connection between the object and either past or anticipated experiences. In the pages ahead, we will see that this supposition can be supported by analysis of what happens in neurological disorders. As is usually the case in mind matters, neurology affords a unique insight into the problem.

DISORDERS OF EXTENDED CONSCIOUSNESS

While loss of core consciousness entails loss of extended consciousness, the converse is not true. Patients in whom extended consciousness is compromised, in one form or another, retain core consciousness. The precedence of core consciousness is thus firmly established.

Transient Global Amnesia

The most astonishing examples of impaired extended consciousness occur acutely and dramatically in a condition known as transient global amnesia. The condition is benign in the sense that patients return to normal. Transient global amnesia can occur in the setting of migraine headaches, sometimes as a prodrome to the headache, sometimes as a substitute for the headache. In transient global amnesia, beginning acutely and lasting for a period of a few hours, usually less than a day, an entirely normal person is suddenly deprived of the records that have been recently added to the autobiographical memory. The mind no longer has available anything having happened in the instants just before, or in the minutes and hours just before. On occasion, nothing that has happened in the days prior to the beginning of the event is available at all.

Considering that our memory of the here and now also includes memories of the events that we constantly anticipate—what I like to call memories of the future—it follows that a person struck by transient global amnesia also does not have available any memory regarding the intended plans for the minutes, hours, or days that lie ahead. It is quite common for the transient global amnesiac to have no inkling whatsoever as to what the future may hold. The person struck by transient global amnesia is thus deprived of both personal historical provenance and personal future but retains core consciousness for the events and objects in the here and now. In effect, when a patient fails to recognize a particular object or person, there is even core consciousness for the fact that some knowledge is no longer present. In spite of adequate consciousness for the current objects and actions, however, the situation fails to make sense to the patient because, without an updated autobiography, the here and now is simply incomprehensible. The predicament of transient global amnesia underscores the significant limitations of core consciousness: Without a provenance for the current placement of objects and a motive for the current actions, the present is nothing but a puzzle. This is probably why, almost invariably, transient global amnesiacs constantly repeat the same anxious questions: Where am I? What am I doing here? How did I come here? What am I supposed to be doing? The patients tend not to ask who they are. They often have a basic sense of their persons, although even that sense is impoverished. If patients with epileptic automatism are good examples of the suspension of core consciousness and of everything that hinges on it—core self, autobiographical self, extended consciousness—patients with transient global amnesia are the perfect example of suspended extended consciousness and autobiographical self, with the preservation of core consciousness and core self.

SOME YEARS AGO we had the opportunity of studying a patient with the mildest of transient amnesia episodes we have ever encountered, and I would like to tell you about her. The patient was a woman of high intelligence and education who led a successful career as an editor. She had a long history of migraine headaches and her health was otherwise excellent. About nine months before her admission to our service she began having classic migraine headaches, sometimes with visual disturbances in one of the visual hemifields and, on occasion, with language difficulties. The headaches had become frequent, one per week. Two weeks before admission, on a routine visit to her family doctor, she complained about her headaches and was referred to our headache clinic with the recommendation that she keep a detailed record of the exact mode of onset, evolution, and possible triggers of her headache. Prior to the event described below, she had recorded details of four different headache episodes as they were happening. Finally, she experienced "a strange event" of which she gave the following account, written in clear handwriting, while her symptoms were occurring. Here is her unedited report.[5]

THURSDAY, AUG 6, 11:05—*At my desk. Suddenly a strange episode. Feel like I'm about to faint or be ill. Vision clear, but whole being concentrates on strange episode. Lean back from my desk. Close eyes. Concentrate on* not *being ill (think of going to rest room—decide against—prefer to sit still). Never lose awareness of surroundings, but am intensely centered on self and odd feeling (never lose sense of where I am or awareness of sounds). Coming out of it feel warm, ask my office mate something about heat in office (by now, five minutes later, don't recall what I said), she indicates it's OK (I think). Now feel right. It's 11:18. But am not quite focused on what I am doing.*

Looking at my work. Don't recognize the page of manuscript I am editing! Flip back and forth, but can't make up my mind what exactly I was doing. (Am clear about the main purpose, not the page I'm on or what I was doing to it.)

Looking at my calendar to enter note of this "event," I find names of people I dealt with in last ten days that disturb me: I am not sure who they are. Most entries, though, are clear to me.

11:23——Reading back. I recall starting to write this but can't recognize the top lines! Feel quite clear-headed now, but still slightly confused about implications, if any? of what I just experienced. At this time head feels clear and OK, maybe a bit heavy. (I'm looking for a headache but it's not there.) I don't dare look at my work to see if it makes better sense than ten minutes ago.

11:25——I read back what I wrote at beginning of the first page: I don't recognize the wording I used! I remember starting to write this, but I'm interested that the beginning of it seems strange.

11:30——Head still clear. No headache. Vision good. Now am trying to recall any relevant circumstances to have on the record. Ordinary morning. Had a cup of coffee at 10 AM. Have been reading and editing a manuscript all morning. Haven't been up from my desk since I got the coffee.

Every time I read back some of what I have written, I find statements that puzzle me because I don't remember putting them down. Trivial wordings, but still they puzzle me because I don't recognize them. (Note: all along, I have been sure of what, who, where I am and what I am doing here.)

11:35——I put the radio on to classical music.

11:45——When I first looked at my calendar to write down the note about this episode, I found I was puzzled by a couple of names I could see. In fact, that's why I started to write this whole account. Now, about half an hour later these names still puzzle me(!). I have looked them up on my department phone list and I can identify who they are and what I did with them, but I am still bothered that the names are strange. The note "both reports on infection control are in" for Aug 3 still is not clear. (I don't recall what occasioned it and this is only Aug 6).

11:50——I think I remember doing the reports, but my mind still can't focus on their content. "Infection Control"?

11:55——*I have remembered where to look to confirm who those names are (but still don't focus on the reports I went over for them). I'm going to lunch.*

12:05——*On my way out, went to rest room then stopped here to reread this and wonder about the significance, besides that it's been a transient episode of some sort. Now out to lunch. Head is slightly heavy all over.*

1:00——*Got to lunch all right. Felt unsure of identity of old friends in the hall. But conversed OK. Got to the lunch line and had moment of panic about how to sign in, then remembered. However, glanced at what person before me wrote on the card to be sure. Started my social security number and had slight panic before finishing it, I assume correctly. Took a healthy lunch, tuna salad and milk. Sat alone. Lingered a bit, thinking of implications of this episode and whether to report it at once to someone? Go home to rest? Ignore it?*

1:20——*Have poured myself a cup of coffee and am getting back to work. Decided at this point to do nothing. Feeling quite stable, unimpaired, and quite sure of what I'm doing (only a bit scared). Poured cup of coffee. Turned on radio to pleasant music, still feel insecure, aware of pulse rate (take it: 80).*

2:05——*Have been working steadily, mostly reviewing this morning's work. Feeling quite normal.*

4:15——*Feeling quite normal. Took a walk at about 4 PM to the library and browsed. Have not tried to read this since 1:20 PM or to test my memory of the items that were unclear earlier in the day.*

5:45——*Before leaving for home, I glance again at the calendar and realize that earlier I was misreading entries on previous days! Now it makes sense, and I remember the reports I worked on and the people involved. Also, I recall that looking over these notes during the afternoon, they seemed different (!) each time I read them. No physical incoordination.*

AUG 7, 10:05 AM——*Woke up fine. Evening OK——head somewhat heavy—— feel anxious——talk to X . . . who suggests sugar problem. Breakfast of two slices of banana nut bread, large chunk of cheese, small orange juice, caffeine-free coffee and ½ t sugar. Went to work. 9:00 AM felt beginning of headache behind eyes (and experienced sweating). 9:30 sure of it: took cup real coffee with two tsp sugar and one tsp sugar out of the spoon. Now, 10:00 AM, head almost clear, but still heavy.*

Have made appointment by phone to discuss work. Have talked business with several persons. OK but maybe my speech expression is slower than usual. I look for my words? Enough of this.

1:25——Later the headache returned. Had lunch at 12:00. Headache never quit. Still basically behind eyes, this time it's the left one and left temple and radiating to back lower left.

AUG 10, 4:30 PM——*Weekend was good. Today good too.*

This unique report was possible because of a number of felicitous circumstances: First, the episode was mild and the patient was less anxious than is usually the case. Second, she had been directed by her physician to write down the precise circumstances in which her headaches occurred and she was thus committed to producing a detailed record of any related event. Lastly, she was an intelligent and cultivated woman who was prepared by personality and even professional training to organize a cogent exposition of her experiences.

The process of core consciousness was maintained throughout the episode, thus permitting her to organize her thoughts and behavior most coherently. Had we been witnesses to the event and had we been interacting with her, I venture that we would have noted something different about her manner, perhaps preoccupation, perhaps vagueness, probably both. But we certainly would have witnessed wakefulness, sustained and focused attention, sustained appropriate behavior, and recognizably motivated emoting. There would be no resemblance whatsoever to the zombie-like behavior of an epileptic during an automatism episode. This is important to note since the acuteness and transiency of the episodes often lead to the unacceptable lumping together of the two conditions. Transient global amnesia and epileptic automatism are as different as night and day.

The transient impoverishment of this patient's autobiographical self, even in the mild form she was fortunate to have, was the dominant manifestation in her condition. The remote biography certainly did exist but the period of time just before the disturbance was missing and even the events of the previous days were retrieved in some sort of penumbra. The diminished availability of biographical information, which was so dramatic for the recent personal experience, was even noticeable in the weak retrieval of identity information. Unable for a moment to retrieve her own name, she nearly panicked.

THE DAYLONG DRAMA of transient global amnesia is frequently telescoped into a matter of less than an hour in the condition of post-traumatic amnesia. Post-traumatic amnesia is a frequent consequence of acute head injuries. A recent patient provided an insightful report: When DT was thrown off his horse and landed on his back, he lost consciousness immediately. The observers who rushed to help him estimate that he remained unconscious for nearly ten minutes. By the time the paramedics arrived, DT had awakened; he looked confused and somewhat agitated and was asking repeatedly about what was going on. His memory for the event begins at about that time and he recalls the unfolding of a clear sequence of states. At first, he looked at the faces peering down at him and could not understand who they were or why they were looking at him. He had no clear idea of who he was, either, and he was even less clear about what he was doing on the ground. Then some sense of who he was came into his mind, although the situation remained inexplicable. A moment later, perhaps after noticing he was wearing his jogging clothes, he announced he wanted to go running; that had indeed been his intention before he had to deal with the misbehaving horse responsible for the whole commotion. It was only by the time he was in the ambulance and on the way to the hospital that a sense of identity began to return.

In less than one hour DT had traversed a variety of neurological conditions. First, a condition not unlike coma or deep dreamless sleep or general anesthesia, in which all forms of consciousness, attention, and wakefulness were suspended. Second, a condition in which wakefulness and minimal attention returned, but core consciousness was still absent, something not unlike certain stages of akinetic mutism or epileptic automatism. Third, a situation not unlike that of transient global amnesia, in which core consciousness had returned but extended consciousness was not present yet. Finally, the entire set of abilities was again made available.

Extended consciousness is also impaired during the progression of Alzheimer's disease. When the loss of memory for past events is marked enough to compromise autobiographical records, the autobiographical self is gradually extinguished and extended consciousness collapses. This happens in advance of the subsequent collapse of core consciousness which I introduced in chapter 3. An event which occurred with the patient and friend I described on page 104 illuminates the problem.

The patient was sitting quietly when he caught sight of his wife as she walked toward him. He showed no sign of recognizing her but returned her warm smile with another warm smile. Knowing that he would not recognize her identity, she said, in her gentle voice, not just, "good morning," but also, "I am your wife." To which he replied, for the first time in the course of the disease: "And who am I?" The question was serious and matter of fact. There was no hint of humor and no anxiety. The inquisitive mode of his former autobiographical self was still in place, as a robust vestige, and it simply wished to know.

The disease had descended from the stage in which learning of new facts and recall of general memories is no longer possible to the stage in which the personal biography can no longer be reliably displayed. Autobiographical self and the extended consciousness that depends on it were now forever gone. Months later it would be time for core consciousness and its simple sense of self to vanish as well.

Anosognosia

Anosognosia provides another good example of impaired extended consciousness without impairment of core consciousness. The word *anosognosia* derives from the Greek *nosos,* "disease," and *gnōsis,* "knowledge," and denotes the inability to recognize a state of disease in one's own organism. Never mind the word should have been "nosoagnosia" rather than *anosognosia* to bring it in line with tradition—think of *prosopagnosia* and *simultanagnosia*—the term stuck.

Neurology has no dearth of bizarre conditions, but anosognosia is one of the strangest. The classical example of anosognosia is that of a victim of stroke, entirely paralyzed in the left side of the body, unable to move hand and arm, leg and foot, face half immobile, unable to stand or walk, who remains oblivious to the entire problem, and who reports that nothing is possibly the matter. When asked how they feel, patients with anosognosia answer with a sincere, "I am fine." This striking condition was first described by Babinski early in the twentieth century.[6]

Those who are fond of "psychological" explanations have long thought that this denial of illness is psychodynamically motivated, that it is nothing but an adaptive reaction to the severe problem the patient faces, colored by the individual's past history relative to comparable situations. They are wrong. It can be easily established that this is not the case by considering the mirror-image situation, that of a patient in whom the right side of the body is paralyzed rather than the left. Those patients do not develop anosognosia. They can be severely paralyzed and even severely aphasic, and yet they are perfectly aware of their tragedy. Anosognosia occurs with right-hemisphere damage. Interestingly, some patients in whom left-side paralysis is caused by a pattern of brain damage different from the one that causes anosognosia can be cognizant of their defects. In short, anosognosia occurs systematically with damage to a particular region of the brain, and only to that region. The denial of illness is caused by the loss of a particular cognitive function, and the cognitive function depends on a particular brain system which is damaged by neurological disease.

The presentation of anosognosia is quite standard. My patient DJ had a complete left-side paralysis, but when I would ask her about her left arm, she would begin by saying that it was fine, that once, perhaps, it had been impaired but not any longer. When I would ask her to move her left arm, she would search around for it and, upon confronting the inert limb, she would ask whether I "really" wanted "it" to move. It was only then, as a result of my insistence, that she would acknowledge that "it doesn't seem to do much by itself." Invariably, she would then have the good hand move the bad arm and state the obvious: "I can move it with my right hand."

This inability to sense the defect automatically, rapidly, and internally through the body's sensory system is nothing less than astounding, while the inability to learn about the defect after repeated confrontation is even more so. Gradually, some patients may recall the many confrontations with the defect and relying on that "externally" obtained information, they may say that they *used* to have that problem, even if they still do.[7]

Patients with anosognosia have damage in the right hemisphere, in a region which includes the cortices in the insula; the cytoarchitectonic areas 3, 1, 2, in the parietal region; and area S_2, also parietal, located in the depth of the sylvian fissure. The damage affects the white matter under these regions, disrupting their interconnection and their connections with the thalamus, the basal ganglia, and the motor and prefrontal cortices. Damage to only parts of this multicomponent system does *not* cause anosognosia. (See figures in the appendix, section 2.)

The brain areas that cross-signal within the overall region of the right hemisphere that is damaged in anosognosia probably produce, through their cooperative interactions, the most comprehensive and integrated map of the current body state available to the brain.[8]

I have suggested that anosognosia results primarily from an inability to represent current body states automatically and through the appropriate signaling channels, which are those of the somatosensory system. In one form or another, this is the most frequent explanation of the problem.[9] But although the traditional explanation may well clarify the main source of the disturbance, we also need to explain why, after patients are specifically told that they are paralyzed, they fail to remember such an important verbal statement a few minutes later. And why, even after they see that they are paralyzed and concur that they are unable to move the left limbs in the same way that they move the right, they also fail to remember such visually presented facts when questioned sometime later. To explain the aspect of anosognosia that allows someone to hold a persistent false belief in spite of having received information to the contrary we need to invoke something more complicated than the mere lack of somatosensory updating. My suggestion is that compromise of somatosensory maps in the right cerebral hemisphere strikes at the heart of the highest level of integrated representation of the organism and, by so doing, undermines part of the biological foundation of the proto-self. The highest level of representation of the organism's current state is no longer comprehensive, and thus no longer available for use in the second-order account of the organism-object relationship, on which consciousness depends. Second-order accounts can still be created out of changes in lower levels of proto-self representation, for instance, in the brain stem. As a consequence, core consciousness is not impaired. But the core self that emerges from it can no longer contribute to autobiographical memory because the contribution to autobiographical memory probably requires the sector of proto-self instantiated at the level of the right somatosensory cortices.

This interpretation only holds when we remember that body representations occur at varied levels, from brain stem to cerebral cortex, and that their contributions vary from level to level. Low-level (brain stem) contributions are essential for the maintenance of core consciousness—other contributions become ineffectual when brain-stem contributions fail; in all probability, high-level (cortex) contributions are mostly necessary to form memories of recent body changes and to update the body component of autobiographical memory.

The lesions that cause anosognosia do not destroy all representations of the organism. They only destroy the set of representations that conjoins, in the greatest detail, the musculoskeletal frame with the state of internal milieu and viscera. The highest level at which this integration can occur is the set of somatosensory maps located in the insula, and in areas S2 and S1 of the right cerebral hemisphere. A number of important organism representations remain intact in anosognosia. They include those in the left cerebral hemisphere homologues of the right insula, and areas S_2 and S_1; in the brain-stem nuclei of the pons and midbrain; and in the hypothalamus. Together those representations provide a partial survey of the organism's state rather than a comprehensive one. Of necessity, they feed autobiographical memory with only partial information, rather than with full-fledged detail.

Anosognosia is a hybrid disorder of consciousness. Patients develop a defect of autobiographical self and their extended consciousness becomes anomalous. In addition, because the lesions also impair the highest-placed components of body representation, the patients have a partially defective proto-self as well.

Asomatognosia

As we have seen, the proto-self depends on varied representations of organism state regarding internal milieu, viscera, vestibular stimulation, and musculoskeletal frame. I am tempted to think that not all of these representations have equal value in the implementation of the proto-self, and I suspect that the internal milieu and the visceral representations are of primary import. A patient, LB, whom I studied some years ago in collaboration with my colleague Steven Anderson, reinforced this idea. The patient had a condition known as *asomatognosia,* which literally means "lack of recognition of the body." Patient LB had sustained a small stroke involving a select part of the right somatosensory cortices. Specifically, the second sensory area (S_2) was damaged. This was not enough to cause any permanent sensory or motor defect, nor for that matter, emotional abnormality. But as can be the case with relatively small vascular lesions, the patient developed seizures arising out of the scarred tissue in her lesion. A remarkable effect was produced in some of the seizure episodes: the patient reported being unable to feel her body, by which she meant, for certain, that she had no awareness of the muscle mass in her limbs and torso. The first time this occurred, the sensation caused alarm. Her mind was working, she knew she was alive and thinking, but she could not feel her body in the usual way. However, she could feel her heart beat, and she decided to administer some "tests" to herself which included pinching her skin and muscle in different parts of the body. At first she felt nothing, but gradually, after several minutes, some sensation came back. After about ten minutes everything was back to normal. Her precise words to describe the episode were "a funny feeling," "like I couldn't feel my body." She was clear about the fact that although this was strange, she was not confused: she knew perfectly well who she was; she knew perfectly well where she was.

After admission to the hospital, and while we attempted to assess her electroencephalographic abnormalities, she was asked to call immediately if any new episode would arise. An episode did arise, a nurse rushed into the room while the episode was unfolding, and we were able to interview her shortly thereafter. The nurse was able to establish that she was oriented to person and place while the episode was in progress. LB was vehement about the fact that she was "alert" and described the situation with amazing precision. "I didn't lose any sense of being, just [lost] my body."

I interpreted the episodes as the result of temporary inactivation of a substantial part of the somatosensory cortical complex within the right hemisphere due to a seizure. The seizure focus was probably located in the border of her S_2 lesion, and the seizure spread into the S_1 region located immediately above in postrolandic gyrus. The highest level of integration for the current state of the organism was temporarily suspended. Nonetheless, the patient continued to have signals about her body available in her brain stem, hypothalamus, in the isolated remnants of her right insula, and in the left somatosensory cortices. Those signals could be transmitted to cingulate cortices. It was mostly the signals pertaining to the musculoskeletal aspect of the body that could not be properly represented in integrated fashion while internal milieu, visceral, and vestibular signaling remained. I presume that internal milieu, visceral and vestibular signaling continued to offer the foundation for her "sense of being" to use her own words. They provided the part of the proto-self on which core consciousness could continue to be generated.

It is important to note that because of the dominance effect of the right somatosensory cortices—they integrate body information for the entire body and thus for both left and right—the defect pertains to both sides of the body even if the lesion is asymmetrically located in the right hemisphere.

The patients with anosognosia, whom we discussed earlier, have far more extensive damage to the right somatosensory cortices, as well as to the underlying connections among them and to connections between them and cingulate cortex, thalamus, and frontal region. Just like patient LB they have core consciousness and are aware of their "being." But the continued defective integration of current signals from the organism leads to a sustained impairment in the updating of autobiographical memory and inevitably disrupts the smooth flow of their conscious minds.

EXTENDED CONSCIOUSNESS IS also compromised in patients who develop major defects of working memory, the most dramatic instances of which occur after extensive frontal lobe damage involving the external aspect of both cerebral hemispheres. The range of images that such patients can hold in mind, at any given time, is quite restricted. Consequently the higher reaches of extended consciousness can no longer be attained.

We can also find examples of impaired extended consciousness in a number of psychiatric conditions, although, given the complexity of these conditions, any interpretation in terms of this framework should be regarded as tentative. Nonetheless, it is reasonable to say that in their acute and severe stages, mania and depression exhibit alterations of extended consciousness. One might venture that the autobiographical self of manic states expands considerably, while the autobiographical self of severe depression shrivels. Some manifestations of schizophrenia, for instance, thought insertion and auditory hallucinations, may be interpreted in part as disorders of extended consciousness. In all likelihood, the patients so affected have anomalous autobiographical memories and deploy anomalous autobiographical selves. It should be noted, however, that during the appearance of such manifestations, the "objects" of their perceptions may be in and of themselves anomalous, and that their proto-selves and core consciousness may be anomalous as well.

Impaired extended consciousness possibly contributes to the dissolution of self associated with states of depersonalization and with states of mystical selflessness, and the same is true of the controversial condition of multiple personalities.

When we discussed core consciousness, I suggested that we consider the behaviors we observe and the conscious mind behind those behaviors on the analogy of an orchestral score with several concurrent parts for the varied groups of musical instruments. I discussed the "behavioral" and "cognitive scores" of persons with impaired or intact core consciousness; I suggest we do the same now for extended consciousness.

The observer of a patient with altered extended consciousness beholds a very different "behavioral score" from that which is produced by a patient with impaired core consciousness. Wakefulness, low-level attention, and background emotions are intact, and so are routine behaviors and some specific emotions. Even simple targeted behaviors can be normally produced. The problem only sets in at the level of highly specific behaviors that depend on substantial knowledge of the past and of the future. Those behaviors are clearly not possible and neither are the emotions related to them.

The "cognitive score" of patients with impaired extended consciousness is a good counterpart to the external observation. The sense of wakefulnesss is present; so is the sense that images are being made and attended; and so is the sense of being alive and capable of feeling. But the higher reaches of meaning are just not available to the personal mind. The mental representation of the autobiographical self is so impoverished that the mind does not know where this self comes from or where it is headed. A life is being sensed but not really examined.

THE TRANSIENT AND THE PERMANENT

The organization of consciousness I propose resolves the apparent paradox identified by William James—that the self in our stream of consciousness changes continuously as it moves forward in time, even as we retain a sense that the self remains the same while our existence continues. The solution comes from the fact that the seemingly changing self and the seemingly permanent self, although closely related, are not one entity but two. The ever-changing self identified by James is the sense of core self. It is not so much that it changes but rather that it is transient, ephemeral, that it needs to be remade and reborn continuously. The sense of self that appears to remain the same is the autobiographical self, because it is based on a repository of memories for fundamental facts in an individual biography that can be partly reactivated and thus provide continuity and seeming permanence in our lives.

This dual arrangement requires the mechanisms of core consciousness and the availability of memory. Core consciousness provides us with a core self, but we also need conventional memory to construct an autobiographical self, and we need both core consciousness and working memory to make the autobiographical self explicit, that is, to display the contents of autobiographical self in extended consciousness. Creatures with limited memory do not face James's paradox. They inhabit a world one step up from innocence. They probably have the seemingly continuous experience of moments of conscious individuality, but they are neither burdened nor enriched by the memories of a personal past, let alone by memories of an anticipated future.

In my proposal, core consciousness is a central resource produced by a circumscribed mental and neural system. The fact that core consciousness is central does not mean that it depends on one structure.

We have already seen that a large number of neural structures is necessary for core consciousness to occur. But the complexity of the system, the multiplicity of its components, and the required concertedness of its operation should not make us overlook the following fact: When we consider the anatomical scale of the whole brain, the basic system underlying core consciousness (the combination of the regions that support the proto-self and of the regions that support the second-order account) is confined to one set of anatomical sites rather than being evenly widespread throughout the brain. There are plenty of brain sites not concerned with the making of core consciousness.

The robustness of core consciousness comes from this anatomical and functional centrality, and from the fact that *any* content of mind, whether actively processed in a live interaction or recalled from memory, can coax the core consciousness system into action, provoke it, so to speak, and in so doing generate a pulse of transient core consciousness. Core consciousness is not organized by sensory modality, say, "visual" core consciousness or "auditory" core consciousness. Rather, core consciousness can be *used* by any sensory modality and by the motor system to generate knowledge about any object or movement.

The contents of the autobiographical self—the organized, reactivated memories of fundamental facts from an individual's biography—are prime beneficiaries of core consciousness. Whenever an object X provokes a pulse of core consciousness and the core self emerges relative to object X, selected sets of facts from the implicit autobiographical self are also consistently activated as explicit memories and provoke pulses of core consciousness of their own.

At any given moment of our sentient lives, then, we generate pulses of core consciousness for one or a few target objects *and for a set of accompanying, reactivated autobiographical memories.* Without such autobiographical memories we would have no sense of past or future, there would be no historical continuity to our persons. But without the narrative of core consciousness and without the transient core self that is born within it, we would have no knowledge whatsoever of the moment, of the memorized past, or of the anticipated future that we also have committed to memory. Core consciousness is a foundational must. It takes precedence, evolutionarily and individually, over the extended consciousness we now have. And yet, without extended consciousness, core consciousness would not have the resonance of past and future. The interlocking of core and extended consciousnesses, of core and autobiographical selves, is complete.

THE NEUROANATOMICAL BASIS FOR THE AUTOBIOGRAPHICAL SELF

To discuss the neuroanatomical basis of the autobiographical self I will invoke the theoretical framework with which I have considered the relation between mental images and the brain. The framework posits an *image space*, the space in which images of all sensory types explicitly occur and which includes the manifest mental contents which core consciousness lets us know, and a *dispositional space*, a space in which dispositional memories contain records of implicit knowledge on the basis of which images can be constructed in recall, movements can be generated, and the processing of images can be facilitated. Dispositions can hold the memory of an image perceived on some previous occasion and can help reconstruct a similar image from that memory; dispositions can also assist the processing of a currently perceived image—for instance, in terms of the degree of attention accorded to the image and the degree of its subsequent enhancement.

There is a neural counterpart of image space and a neural counterpart of dispositional space. Structures such as the early sensory cortices of varied modalities support neural patterns that are likely to be the basis for mental images. On the other hand, higher-order cortices and varied subcortical nuclei hold dispositions with which both images and actions can be generated, rather than holding or displaying the explicit patterns manifest in images or actions themselves. (See figure A.5 in the appendix, for placement of early sensory cortices and higher-order cortices.) I have proposed that dispositions are held in neuron ensembles known as *convergence zones.*[10] To the partition of cognition between an image space and a dispositional space, then, corresponds a partition of the brain into (1) neural-pattern maps, activated in the early sensory cortices, the so-called limbic cortices, and some subcortical nuclei, and (2) convergence zones, located in the higher-order cortices and in some subcortical nuclei. (See appendix, section 3 for further discussion of this issue.)

THE BRAIN FORMS memories in a highly distributed manner. Take, for instance, the memory of a hammer. There is no single place of our brain where we will find an entry with the word *hammer* followed by a neat dictionary definition of what a hammer is.[11] Instead, as current evidence suggests, there are a number of records in our brain that correspond to different aspects of our past interaction with hammers: their shape, the typical movement with which we use them, the hand shape and the hand motion required to manipulate the hammer, the result of the action, the word that designates it in whatever many languages we know. These records are dormant, dispositional, and implicit, and they are based on separate neural sites located in separate high-order cortices. The separation is imposed by the design of the brain and by the physical nature of our environment. Appreciating the shape of a hammer visually is different from appreciating its shape by touch; the pattern we use to move the hammer cannot be stored in the same cortex that stores the pattern of its movement as we see it; the phonemes with which we make the word *hammer* cannot be stored in the same place, either. The spatial separation of the records poses no problem, as it turns out, because when all the records are made explicit in image form they are exhibited in only a few sites and are coordinated in time in such a fashion that all the recorded components appear seamlessly integrated.

If I give you the word *hammer* and ask you to tell me what "hammer" means, you come up with a workable definition of the thing, without any difficulty, in no time at all. One basis for the definition is the rapid deployment of a number of explicit mental patterns concerning these varied aspects. Although the memory of separate aspects of our interaction with hammers are kept in separate parts of the brain, in dormant fashion, those different parts are coordinated in terms of their circuitries such that the dormant and implicit records can be turned into explicit albeit sketchy images, rapidly and in close temporal proximity. The availability of all those images allows us, in turn, to create a verbal description of the entity and that serves as a base for the definition.

I would like to suggest that the memories for the entities and events that constitute our present autobiography are likely to use the same sort of framework used for the memories we form about any entity or event. What distinguishes those memories is that they refer to established, invariant facts of our personal histories.

I propose we store records of our personal experiences in the same distributed manner, in as varied higher-order cortices as needed to match the variety of our live interactions. Those records are closely co-ordinated by neural connections so that the contents of the records can be recalled and made explicit, as ensembles, rapidly and efficiently.

The key elements of our autobiography that need to be reliably activated in a nearly permanent fashion are those that correspond to our identity, to our recent experiences, and to the experiences that we anticipate, especially those in the near future. I propose that those critical elements arise from a continuously reactivated network based on convergence zones which are located in the temporal and the frontal higher-order cortices, as well as in subcortical nuclei such as those in the amygdala. The coordinated activation of this multisite network is paced by thalamic nuclei, while the holding of the reiterated components for extended periods of time requires the support of prefrontal cortices involved in working memory. In short, the autobiographical self is a process of coordinated activation and display of personal memories, based on a multisite network. The images which represent those memories explicitly are exhibited in multiple early cortices. Finally, they are held over time by working memory. They are treated as any other objects are and become known to the simple core self by generating their own pulses of core consciousness.

The sustained display of autobiographical self is the key to extended consciousness. Extended consciousness occurs when working memory holds in place, simultaneously, *both* a particular object *and* the autobiographical self, in other words, when *both* a particular object *and* the objects in one's autobiography simultaneously generate core consciousness.

THE AUTOBIOGRAPHICAL SELF, IDENTITY, AND PERSONHOOD

I have indicated that identity and personhood, the two notions that first come to mind when we think of the word *self,* require autobiographical memory and its actualization in the autobiographical self. The repository of records in autobiographical memory contains the memories that constitute identity along with the memories that help define our personhood. What we usually describe as a "personality" depends on multiple contributions. One important contribution comes from "traits," whose ensemble is often referred to as "temperament," and which are already detectable around the time of birth. Some of those traits are genetically transmitted and some are shaped by early developmental factors. Another important contribution comes from the unique interactions that a growing, living organism engages in a particular environment, physically, humanly, and culturally speaking. This latter contribution—which is made under the continuous shadow of the former—is recorded in autobiographical memory and is the footing for autobiographical self and personhood. In a large array of situations, from simple to complex, from benign to dangerous, involving anything from trivial preferences to ethical principles, the existence of autobiographical memory permits organisms to make generally consistent evocations of emotional as well as intellectual responses.

When we talk of the molding of a person by education and culture, we are referring to the combined contributions (1) of genetically transmitted "traits" and "dispositions," (2) of "dispositions" acquired early in development under the dual influences of genes and environment, and (3) of unique personal episodes, lived under the shadow of the former two, sedimented and continuously reclassified in autobiographical memory. We can imagine the neural counterpart to this complicated process as consisting of the creation of dispositional records on the basis of which the brain can evoke, given the appropriate stimulus, a collection of fairly simultaneous responses ranging from emotions to intellectual facts. Using the convergence-zone framework, we can imagine that these responses are controlled by records in particular brain sites which direct the playing out of the responses in a variety of structures—early sensory cortices for the depiction of sensory images of varied nature; motor and limbic cortices and subcortical nuclei for the execution of a large range of actions including those that constitute emotions.

Not only are there many such convergence zones/disposition sites, but they are not even contiguously located. In all likelihood, some are located in the cortex while others are in subcortical nuclei. Those in the cortex are distributed in the temporal as well as the frontal regions. In those personalities that appear to us as most harmonious and mature from the point of view of their standard responses, I imagine that the multiple control sites are interconnected so that responses can be organized, at varied degrees of complexity, some involving the recruitment of just a few brain sites, others requiring a concerted large-scale operation, but often involving both cortical and subcortical sites.

The simple notion of identity is derived from precisely this arrangement. In a number of sites of both temporal and frontal regions, convergence zones support dispositions that can consistently and iteratively activate, within early sensory cortices, the fundamental data that define our personal and social identities—everything from the fabric of our kinships, to the network of our friends, to the roster of places that have marked our lives, all the way to our given names. Our identities are displayed in sensory cortices, so to speak. At any moment of our waking and conscious lives, a consistent set of identity records is being made explicit in such a way that it forms a backdrop for our minds and can be moved to the foreground rapidly if the need arises. Under some circumstances the range of activated records can be enlarged to include a greater sweep of our personal histories and of our anticipated futures. But moment by moment, whether or not we enlarge the scope of such memories, they are active and available. We know that their inactivation does not go by unnoticed—the result of their inactivation is some variant of transient global amnesia.

When I first thought of this explanation for the process behind our sense of identity, I wondered about the burden of constant repetition and internal reenactment of the very same sensory patterns in order to display the very same information. Would this not be an intolerable burden for neurons? But I felt reassured when I thought of other examples of seemingly inordinate burdens on biological tissue. Think of the muscle cells in your heart sentenced for life to their repeated contraction.

THE IDEA EACH of us constructs of ourself, the image we gradually build of who we are physically and mentally, of where we fit socially, is based on autobiographical memory over years of experience and is constantly subject to remodeling. I believe that much of the building occurs nonconsciously and that so does the remodeling (see section ahead on the unconscious). Those conscious and unconscious processes, in whatever proportion, are influenced by all sorts of factors: innate and acquired personality traits, intelligence, knowledge, social and cultural environment. The autobiographical self we display in our minds, at this moment, is the end product not just of our innate biases and actual life experiences, but of the reworking of memories of those experiences under the influence of those factors.

The changes which occur in the autobiographical self over an individual lifetime are not due only to the remodeling of the lived past that takes place consciously and unconsciously, but also to the laying down and remodeling of the anticipated future. I believe that a key aspect of self evolution concerns the balance of two influences: the lived past and the anticipated future. Personal maturity means that memories of the future we anticipate for the time that may lie ahead carry a large weight in the autobiographical self of each moment. The memories of the scenarios that we conceive as desires, wishes, goals, and obligations exert a pull on the self of each moment. No doubt they also play a part in the remodeling of the lived past, consciously and unconsciously, and in the creation of the person we conceive ourselves to be, moment by moment.

Our attitudes and our choices are, in no small part, the consequence of the "occasion of personhood" that organisms concoct on the fly of each instant. Little wonder, then, that we can vary and waver, succumb to vanity and betray, be malleable and voluble. The potential to create our own Hamlets, Iagos, and Falstaffs is inside each of us. Under the right circumstances, aspects of those characters can emerge, briefly and transiently, one hopes. In some respects, it is almost astonishing that most of us have only *one* character, although there are sound reasons for the singularity. The tendency toward unified control prevails during our developmental history, probably because a single organism requires that there be one single self if the job of maintaining life is to be accomplished successfully—more than one self per organism is not a good recipe for survival. The rich imaginings of our mind do prepare "multiple drafts" for our organism's life script—to place the idea in the framework proposed by Daniel Dennett.[12] Yet, the shadows of the deeply biological core self and of the autobiographical self that grows under its influence constantly propitiate the selection of "drafts" that accord with a single unified self. Moreover, the delicately shaped selectional machinery of our imagination stakes the probabilities of selection toward the same, historically continuous self. We can be Hamlet for a week, or Falstaff for an evening, but we tend to return to home base. If we have the genius of Shakespeare, we can use the inner battles of the self to create the entire cast of characters in Western theater—or, in the case of Fernando Pessoa, to create four distinct poets under the same pen. But, at the end of it all, it is the selfsame Shakespeare who retires quietly to Stratford, and the selfsame Pessoa who drinks himself to oblivion in a Lisbon hospital. In short, there are limits to the unified, continuous, single self, as Whitehead points out in his comments on self-consciousness in *Process and Reality*[13]; human failings and the strange condition of multiple personalities testify to the existence of such limits; and yet the tendency toward one single self and its advantage to the healthy mind are undeniable.[14]

THE AUTOBIOGRAPHICAL SELF AND THE UNCONSCIOUS

Florestan, the romantic hero of Beethoven's *Fidelio*, is unjustly imprisoned in a dark dungeon. "God, it is dark in here!" he exclaims, and he might as well be referring to the darkness at the bottom of human memory.[15] We are not conscious of which memories we store and which memories we do not; of how we store memories; of how we classify and organize them; of how we interrelate memories of varied sensory types, different topics, and different emotional significance. We have usually little direct control over the "strength" of memories or over the ease or difficulty with which they will be retrieved in recall. We have all sorts of interesting intuitions, of course, about the emotional value, the robustness, and the depth of memories but *not* direct knowledge of the mechanics of memory. We have a solid corpus of research on factors governing learning and retrieval of memory, as well as on the neural systems required to support and retrieve memories.[16] But direct, conscious knowledge, we do not have.

The memories which constitute our autobiographical records are in precisely these same circumstances, perhaps all the more so because the high emotional charge of so many of those memories may lead the brain to treat them differently. We experience the contents that go into the autobiographical records—we are conscious of those contents—but we know not how they get stored; how much of each; how robustly; how deeply or how lightly. Nor do we know how the contents become interrelated as memories and are classified and reorganized in the well of memory; how linkages among memories are established and maintained over time, in the dormant, implicit, dispositional mode in which knowledge exists within us. And yet, while we do not experience any of this directly we do know a little about the circuits that hold those memories. They are abundantly located in higher-order cortices, especially those of the temporal and frontal regions, and hold close network relations with cortical and subcortical limbic regions and with the thalamus. Neurobiologically speaking, Florestan's dark dungeon will get some light before too long.

To be sure, certain sets of autobiographical memories are simply and consistently reactivated moment by moment, and these memories deliver to our extended consciousness the facts of our physical, mental, and demographic identity; the facts of our recent provenance (where we were just before, a few minutes and hours ago, the day before) and the facts of our intended immediate future (what we must accomplish over the next minutes and hours, where we are headed tonight and tomorrow). Disruption of this fundamental aspect of autobiographical self causes the dramatic neurologic problem we encountered in transient global amnesia.

Certain contents of autobiographical memory, however, remain submerged for long periods of time and may always remain so. It is easy to imagine, given that memories are not stored in facsimile fashion and must undergo a complex process of reconstruction during retrieval, that the memories of some autobiographical events may not be fully reconstructed, may be reconstructed in ways that differ from the original, or may never again see the light of consciousness. Instead, they may promote the retrieval of other memories which do become conscious in the form of other concrete facts or as concrete emotional states. In the extended consciousness of that moment, the facts so retrieved may be unexplainable because of their apparent lack of connection with the contents of consciousness that command center stage then. The facts may appear unmotivated, although a web of connections does indeed exist sub rosa, reflecting either the reality of some moment lived in the past or the remodeling of such a moment by gradual and unconscious organization of covert memory stores.

Now consider the multiple and legitimate meanings of the word *connections* in the previous sentence. The word refers to the connection of things and events as it may have occurred historically; it refers to the mental imagetic representation of those things and events as we experience them; and it also refers to the neural connection among brain circuits necessary to hold the record of things and events and redeploy such records in explicit neural patterns. The world of the psychoanalytic unconscious has its roots in the neural systems which support autobiographical memory, and psychoanalysis is usually seen as a means to see into the tangled web of psychological connections within autobiographical memory. Inevitably, however, that world is also related to the other kinds of connections I just outlined.

The unconscious, in the narrow meaning in which the word has been etched in our culture, is only a part of the vast amount of processes and contents that remain nonconscious, not known in core or extended consciousness. In fact, the list of the "not-known" is astounding. Consider what it includes:

1. all the fully formed images to which we do not attend;
2. all the neural patterns that never become images;
3. all the dispositions that were acquired through experience, lie dormant, and may never become an explicit neural pattern;
4. all the quiet remodeling of such dispositions and all their quiet renetworking—that may never become explicitly known; and
5. all the hidden wisdom and know-how that nature embodied in innate, homeostatic dispositions.

Amazing, indeed, how little we ever know.

NATURE'S SELF AND CULTURE'S SELF

It is usually foolhardy to revisit the nature versus nurture argument and try to decide whether a certain cognitive function is shaped in a particular manner and in a particular individual by the genome, via its related biological constraints, or by the environment, via the influence of culture. Curiously, when we look at consciousness in the perspective of my proposal, distinctions of this sort appear somewhat more possible. For instance, I would venture that virtually all of the machinery behind core consciousness and the generation of core self is under strong gene control. Barring situations in which disease disrupts brain structure early on, the genome puts in place the appropriate body-brain linkages, both neural and humoral; lays down the requisite circuits, and, with help from the environment, allows the machinery to perform in reliable fashion for an entire lifetime.

The development of the autobiographical self is a different matter. To be sure, the connection between core self and the structures which support the development of autobiographical memory is organized under genomic control. So are the processes on the basis of which learning can take place and modeling of cortical and subcortical circuits can occur so that convergence zones and their dispositions are put in place. In other words, autobiographical memory develops and matures under the looming shadow of an inherited biology. However, unlike the core self, much will occur in the development and maturation of autobiographical memory that is not just dependent on, but is even regulated by, the environment. For instance, the schedules of reward and punishment offered to developing infants, children, and adolescents do vary among different home, school, and social environments; the shaping of the events which constitute the historical past of an individual and his or her anticipated future is controlled in no small measure by the environment; the rules and principles of behavior governing the cultures in which an autobiographical self is developing are under the control of the environment; likewise for the knowledge according to which individuals organize their autobiography, which ranges from the models of individual behavior to the facts of a culture.

When we talk about the self in order to refer to the unique dignity of a human being, when we talk about the self to refer to the places and people that shaped our lives and that we describe as belonging to us and as living in us, we are talking, of course, about the autobiographical self.

The autobiographical self is the brain state for which the cultural history of humanity most counts.

BEYOND EXTENDED CONSCIOUSNESS

Extended consciousness allows human organisms to reach the very peak of their mental abilities. Consider some of those: the ability to create helpful artifacts; the ability to consider the mind of the other; the ability to sense the minds of the collective; the ability to suffer with pain as opposed to just feel pain and react to it; the ability to sense the possibility of death in the self and in the other; the ability to value life; the ability to construct a sense of good and of evil distinct from pleasure and pain; the ability to take into account the interests of the other and of the collective; the ability to sense beauty as opposed to just feeling pleasure; the ability to sense a discord of feelings and later a discord of abstract ideas, which is the source of the sense of truth. Among this remarkable collection of abilities allowed by extended consciousness, two in particular deserve to be highlighted: first, the ability to rise above the dictates of advantage and disadvantage imposed by survival-related dispositions and, second, the critical detection of discords that leads to a search for truth and a desire to build norms and ideals for behavior and for the analyses of facts. These two abilities are not only my best candidates for the pinnacle of human distinctiveness, but they are also those which permit the truly human function that is so perfectly captured by the single word *conscience*. I do not place consciousness, either in its core or extended levels, at the pinnacle of human qualities. Consciousness is necessary, but not sufficient, to reach the current pinnacle.

The enchainment of precedences is most curious: the nonconscious neural signaling of an individual organism begets the *proto-self* which permits *core self* and *core consciousness*, which allow for an *autobiographical self*, which permits *extended consciousness*. At the end of the chain, *extended consciousness* permits *conscience*.

The status of our understanding relative to conscience, extended consciousness, and core consciousness may well parallel the order in which humans seem to have realized the existence of such phenomena and become curious about them. Humans had identified conscience and had an interest in its doings long before they identified extended consciousness as a problem, let alone core consciousness. The gods of antiquity do not speak to the heroes of the Homeric poems about matters of consciousness but rather about matters of conscience: think of Athena when she restrains the arm of the boy Achilles and stops him from killing Agamemnon in the *Iliad*. Ten centuries B.C., the Homeric tales presume the existence of core consciousness but never dwell on it explicitly. They describe, indirectly, a patchy, god-ridden consciousness, but what they really worry about is conscience.[17] Solon, seven centuries B.C., is probably on the path to both conscience and consciousness—he does advise the reader "to know thyself."[18] Similarly wise are the Greeks from 500 B.C. onward, as well as the writers and protagonists of Genesis, the authors of the Mahābhārata, and the *shi* who collected the Tao-te Ching. But none of them deal with the notions of consciousness that preoccupy us now. It is not just that the word for consciousness is not to be found in Plato or Aristotle, neither *nous* nor *psyche* being equivalents. The concept is not there, either. (*Psyche* did refer to some aspects of an organism that I believe are critical for the appearance of what we now call consciousness [breath; blood] or that are closely related [mind, soul], but it did not correspond to the same concept.[19]) The preoccupation with what we call consciousness now is recent—three and a half centuries perhaps—and has only come to the fore late in the twentieth century.

The coinage of the words by which we denote the "phenomena of consciousness" in the languages that carried Western thought to us also suggests that curiosity about and the understanding of these phenomena probably marched in the reverse order of their complexity. In the history of the English language, for instance, the Middle English word related to consciousness is *inwit,* a superb construction blending the notion of interior (*in*) with that of mind (*wit*). The word *conscience* (from the Latin *con* and *scientia,* which suggests the gathering of knowledge) has been in usage since the thirteenth century, while the words *consciousness* and *conscious* only appear in the first half of the seventeenth century, well after the death of Shakespeare (the first recorded usage of the word *consciousness* dates to 1632). By 1600, Shakespeare had Hamlet say, "Thus conscience does make cowards of us all," and he really meant conscience, not consciousness. Although the bard understood deeply the nature of extended consciousness and virtually planted it in literary form within Western culture, he could never name it as such. He may even have realized that something like core consciousness lurked behind extended consciousness, but core consciousness was not a focus of his concerns.

In English and in its "mother language" German, there are separate words for conscience and consciousness. In German the word for "consciousness" is *Bewusstsein,* and the word for "conscience" is *Gewissen.* In Romance languages, however, one single word denotes both conscience and consciousness. When I translate the word "unconscious" in French (*inconscient*) or in Portuguese (*inconsciente*), I can be referring either to a person who is comatose or to a person whose behavior is unconscionable. English offers us both *unconscious* and *unconscionable,* and German gives us *unbewusst* and *gewissenlos.* But in Romance languages, only the context can reveal which meaning is meant. On this issue, incidentally, matters can get quite confusing but always interesting. In Romance languages such as French or Portuguese, we can also refer to consciousness with a word denoting knowledge, e.g., in French *connaissance,* in Portuguese *conhecimento.* Note that, once again, the alternate word refers to "facts known," presumably the facts that there is a self and that there is knowledge attributed to it. Whatever the word for consciousness, we are never far from the notion of encompassing knowledge, as betrayed by some variation on *con* (an embracing with) and *scientia* (facts, scientific and otherwise).

When the concept behind the word *consciousness* began to emerge, users of Romance languages recruited the word *conscience* to denote it rather than coining a new word. The cultural tolerance of the conflation of meanings is most curious, perhaps another testimony to the evolution of human concerns on these matters, and is worth investigating in its own right. Somehow, the relatedness between the concepts of conscience and consciousness was given more value than their distinctiveness. Curiously, unlike English or German, Romance languages also have no word for *self* (the reflexive pronouns are not a good substitute). The personal pronouns *I* or *me* are deemed satisfactory to designate an entity that could have its own name—a direct translation of self—but does not.

One might have guessed that because conscience is at the top of the complexity heap I have just outlined, it would have been the last phenomenon to be considered and understood in terms of its nature and mechanisms. The opposite seems to be true. I would argue that we know more about the workings of conscience than we do about those of extended consciousness, in the same way that we know more about extended consciousness than we do about core consciousness. The work of Jean-Pierre Changeux on the neurobiology of ethics, or of Robert Ornstein on the relation between consciousness and society, supports my contention on conscience. Daniel Dennett's, Bernard Baars's, and James Newman's attempts to elucidate consciousness at the level of extended consciousness support the latter contention.[20] As far as I can see, the balance of the mystery lies behind core consciousness. It may well be that conscience and extended consciousness are incompletely explained only because understanding them depends in part on resolving the problem of core consciousness.

Chapter Eight

The Neurology
of Consciousness

I regard the proposal outlined in the previous chapters as ground zero
for a research program on the neural basis of consciousness. Only future investigation of these proposals, using a variety of approaches,
will decide the merits of the ideas presented here. In the meantime,
however, we can consider these ideas in terms of the evidence already
available in neuroscience and that is the purpose of this chapter.

In chapters 5, 6, and 7, I advanced hypotheses regarding mechanisms
for core consciousness and extended consciousness, and indicated
which anatomical structures were necessary to support the proto-self
and the second-order map required by those mechanisms.

Based on those hypotheses, the following statements should be true:

1. Bilateral damage to maps of somatosensory information, which
    form the neural basis for the proto-self, should disrupt con-

sciousness. The disruption of consciousness should be maximal following damage at the level of the upper brain stem and hypothalamus, where proto-self structures are tightly packed together, and less severe at higher levels (the cortices of insula, S2, S1; related parietal association cortices), where processing chains are spatially more separated.

2. Bilateral damage to structures presumed to participate in constructing the second-order imaged account of the organism-object relationship should disrupt core consciousness partially or completely. Examples of such structures are certain nuclei of the thalamus and the cingulate cortices.

3. Bilateral damage to temporal cortices, including the inferotemporal region known as IT and the temporal pole known as TP, should not impair core consciousness, since in those circumstances the structures required to represent the proto-self, to process most objects to be known, and to create the imaged account of the organism-object relationship are all intact. However, damage to the temporal cortices will impair the activation of autobiographical memory records and thus reduce the scope of extended consciousness. The same applies to bilateral damage in some higher-order cortices within the vast prefrontal regions, which also support the records from which the autobiographical self can be activated.

4. Bilateral damage to the hippocampus will not impair core consciousness. However, because new learning of facts will be precluded, it will halt the growth of autobiographical memory, affect its maintenance, and, consequently, alter the quality of extended consciousness in the future.

5. Bilateral damage to early sensory cortices concerned with external sensory information (e.g., vision, hearing) should not impair core consciousness except by precluding the representation of the aspects of a given object which depend on that particular cortex. The situation of somatosensory cortices is exceptional since they provide part of

the basis for the proto-self. Their damage is referred to in statement 1 above.

6. Bilateral damage to prefrontal cortices, even if extensive, should not alter core consciousness.

In the pages ahead I assess the validity of these statements in the light of evidence from neuropathology, neuroanatomy, neurophysiology, and neuropsychology.

Assessing Statement Number One: Evidence for a Role of Proto-Self Structures in Consciousness

Statement number one indicates that bilateral damage to maps of somatosensory information which form the neural basis for the proto-self should disrupt consciousness. This statement is supported by a combination of evidence from cases of coma, persistent vegetative state, locked-in syndrome, and basal-forebrain damage. Because the amount of evidence is so vast, I will focus on material pertaining to coma and persistent vegetative state and begin by offering a brief description of what coma and vegetative state look like.[1]

It Looks like Sleep

It looks like sleep, it may sound like sleep, but it is not sleep. There is a universal history for the presentation of coma, and the clinical description is likely to read as follows: Without any warning, the patient collapsed, was suddenly on the ground, and was breathing with some difficulty; he never responded to his wife or to the paramedics when they came to take him to the hospital; he never responded to anyone in the emergency room; he still did not respond to the physicians four days later. If it were not for the elaborate wiring and tubing and digital displays surrounding him, if it were not for the fact that this is a high-tech unit for treating cerebrovascular diseases, you, as a visitor, might think that he is simply asleep. But, the fact is, he has indeed had a stroke and is in a coma, a very abnormal state from which no amount of regular stimulation will awaken him.

You can talk to him, you can whisper in his ear, you can touch his face or squeeze his hand, you can perform all the manipulations required to evaluate such situations, but he won't wake up. And yet his heart is beating, his blood is circulating, his lungs are breathing, his kidneys are working, too, and so are other organs and systems required for immediate survival, with a little help from the intensive-care team. The brain is the problem. It has been damaged by a stroke in a small but critical region. The observable result is a suspension of wakefulness, emotion, attention, purposeful behavior. The result you could infer from your observation is that consciousness has been suspended as well. Not only is he unable to report any evidence of a conscious mind at work, but he gives none of the indirect signs that he might have one. He is alive and yet his organism has changed radically.

Every night, when we fall asleep and reach the deep, refreshing, dreamless stage of sleep known as stage 4, we are, in terms of consciousness and of mind, in a state similar to his. We count on waking up again and so we have no anxiety about giving up consciousness and mind for a few hours. The condition of the comatose patient is quite different, however: he cannot be awakened from the kind of sleep state he has been forced in, and the probability of his recovering consciousness is not high. It is possible that his coma will persist and that death will eventually ensue. It is also possible that his deep coma will become lighter and eventually turn into a permanent state of unconsciousness known as persistent vegetative state.

If the condition evolves into vegetative state, the patient will begin showing cycles of apparent sleep and wakefulness, which will succeed each other in a seemingly normal way. This is something you can tell from two sources of evidence. First, his electroencephalogram (EEG) will change and may show, during a certain number of hours of each day, the patterns characteristic of sleep or wakefulness. Second, he may begin responding to stimuli by opening his eyes. Unfortunately, neither piece of evidence indicates that consciousness is returning; all it indicates is that wakefulness has returned. As we discussed, wakefulness is a necessary element in consciousness (dreams excepted, of course), but by no means is it the same as consciousness. If the patient becomes vegetative, his control of autonomic functions such as blood pressure and breathing may also normalize. Otherwise, in rare patients and on rare occasions, there may be isolated instances of coordinated movements of head and eyes, isolated stereotypical utterances, an isolated smile or tear. In essence, however, during the seemingly wakeful part of a day, patients in vegetative state have no behavior whatsoever, neither spontaneously nor in response to a prompt, that betrays the presence of consciousness. Emotion, attention, and purposeful behavior do not return in the vegetative state. The reasonable assumption, which is corroborated by the reports of rare individuals in whom consciousness did return eventually, is that consciousness is still out of the picture.[2]

The cause of this patient's tragedy is damage to a minuscule part of the brain stem. The brain stem connects the spinal cord to the large expanses of the cerebral hemispheres. It is the tree-trunk-like structure that links the part of the central nervous system which sits inside the vertebral canal, up and down the spine—the spinal cord—to the part of the central nervous system which sits inside the skull—the brain, in the usual sense. The brain stem receives signals from the entire body proper and also serves as conduit for those signals as they travel toward parts of the brain situated higher up; likewise it serves as conduit for signals traveling in the opposite direction from the brain toward the body proper. In addition, it holds numerous small nuclei and local interconnecting nerve fibers. It has long been known that the control of life functions, such as those of the heart and lungs and gut, depends on the brain stem, as does the control of sleep and wakefulness. Thus, in an extremely small area of brain, nature tightly packs many of the critical pathways which signal chemical and neural events from the body proper to the central nervous system and which bring signals from the central nervous system to the body proper.

Alongside those critical pathways, there are myriad tiny centers which control many life operations.

None of these pathways or control centers is strewn about randomly. On the contrary, as is always the case in the brain, they are arranged in consistent anatomical patterns which can be found in all humans, in exactly the same arrangement, and can be found in many other species in almost the same position.[3] When coma occurs as a result of damage below the level of the thalamus, the destruction occurs from the level of the mid to upper pons on upward toward the midbrain and hypothalamus. Moreover, the damage must be located in the back part of the brain stem rather than the front.[4]

The damage that causes coma and persistent vegetative state tends to spare several cranial nerve nuclei and several long descending and ascending tracts, but it consistently injures several families of nuclei in the brain-stem tegmentum. These include such well-known reticular nuclei as the cuneiform nucleus and the nucleus pontis oralis. I will refer to such nuclei as *classical reticular nuclei.* But the damage also encompasses "nonclassical" nuclei which, depending on the author, may or may not be lumped together under the somewhat controversial designation of "reticular formation." Those nonclassical nuclei include a collection of monoamine nuclei (locus coeruleus, ventral tegmental area, substantia nigra, raphe nuclei), acetylcholine nuclei, and sizable aggregates of nuclei known as the parabrachial nuclei and the periaqueductal gray. Finally, the colliculi may be damaged also, but whether or not they are, their inputs and outputs are severed. Their functions are either compromised or, if not, the results of those functions cannot be delivered to either the brain stem or telencephalon. (See figure 8.1. The reticular formation is marked in the shaded area.)

Are the situations of coma and persistent vegetative state supportive of statement number one? I believe they are, although several comments are in order at this point. As noted, the extent of brain-stem damage that usually causes coma compromises many structures, from those in the classical reticular nuclei, which are known to

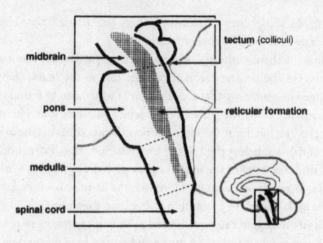

Figure 8.1. Main anatomical divisions of the brain stem, seen in a sagittal section through the brain's midline. The anatomical orientation is shown in the inset panel located to the right of the main panel.

control wakefulness, to the nonclassical nuclei, which easily fit the notion of proto-self I advanced. It might be argued that the impairment of consciousness seen in coma is parsimoniously explained by damage to classical reticular nuclei. Leaving aside the fact that the neuropathological and neuroanatomical evidence in these cases is not complete yet, the argument would be problematic because the likelihood of the distinct but contiguous families of nuclei having truly independent functions is low. The argument would overlook the anatomical placement and functional neighborhood of the classical reticular nuclei and of the monoamine/acetylcholine nuclei. Those nuclei are anatomically and functionally interwoven with those that regulate current body state and that map body state, and it is apparent that reticular and monoamine/acetylcholine nuclei are influenced by events in body-related nuclei.[5] I am not suggesting that classical reticular and monoamine/acetylcholine nuclei do not do what they have been presumed to do—activate and modulate the thalamus and the cerebral cortex. But I am suggesting that they do so under circumstances that are set, in good part, by the proto-self structures which regulate the body and represent the body state in the brain stem. We need to include the body-regulating structures in the picture we draw of the consciousness-related brain stem, and perhaps we also need to widen the anatomical characterization of the proto-self and include the classical reticular nuclei — this is a matter for future research and cannot be decided now.

One other reason why the argument would not be valid pertains to the fact that some comatose patients, with no sign whatever of consciousness, may have a normal electroencephalogram, which might indicate that the functions of the classical reticular nuclei are somehow preserved (or quite simply, that we must be cautious about the interpretation of EEG findings relative to consciousness since it is also true that conscious patients may have an abnormal EEG).[6]

In some instances, coma occurs following combined damage to the upper midbrain and the hypothalamus or following damage to the thalamus. In both of these instances, the situation is also compatible with statement number one. Damage to the upper midbrain and hypothalamus impairs a sizable portion of the structures required to implement the proto-self. No less importantly, the damage stands in the way of pathways ascending toward cortical proto-self and second-order map sites. The same reasoning applies to instances of damage to the thalamus.

Importantly, in cases of brain-stem damage in which consciousness is not impaired, for instance, in locked-in syndrome, the region described above remains intact: nearly all the structures I just enumerated are outside the area damaged in locked-in. The very different presentation of locked-in deserves a special comment.

It May Look like Coma

If the brain-stem lesions that cause coma can help us assess statement number one, so can the lesions that do not cause coma, especially when they are located in close proximity to those that do. The most striking example occurs when damage to a region of the brain stem only a few millimeters away from the region I just described for coma produces, instead, a devastating condition known as locked-in syndrome. As I indicated in the chapter on emotion, patients with locked-in syndrome lose their ability to move voluntarily but remain conscious. Let me give you an idea of the situation.

Just as with coma, the tragedy will usually have begun without warning. The patient will be on the floor just as suddenly as the coma patient, motionless and speechless, and motionless and speechless she will remain after the terrible event, for as long as she lives. Everyone around her will have suspected a stroke and, at first, for a period of hours, days, or weeks, she will have lapsed into a coma. But sooner or later, at some point in the course of the hospital admission, it will become apparent that although motionless, the patient is awake. Someone will suspect that she is probably conscious. There will have been few clues: the eyes and perhaps the sense that a keen observer will have had that the patient blinked meaningfully. In the blink of an eye, the fate of the patient will have changed. Upon careful examination it will be found that she can still perform one type of movement: she can move her eyes up and down, and she can blink. She cannot frown, she cannot move her eyes sideways, she cannot move her lips or stick out her tongue, and she cannot move her neck, arms, or legs. Eye movements in the vertical direction and blinking are all that remain of the ability to act under willful control. Because of these modest residual abilities, she can be asked to move her eyes up and she does do so immediately, and likewise she can move them down upon request. She can clearly hear us speaking and she can understand the meaning of our words. She is conscious. She is not in a coma. Her situation is known as locked-in syndrome, an apt description for the state of near-solitary confinement of the patient's mind.

The simple motor ability that remains permits an emergency communication code: The patient can be asked to signify yes by moving her eyes up and to signify no by moving her eyes down. And her blinking can be used to detect a letter of the alphabet out of a list recited to her, such that she can compose words and sentences, letter by letter, and thus communicate complicated thoughts. These codes allow patients to answer questions pertaining to their history and current state, and allow nurses, physicians, and family to maintain a helpful dialogue. Coma is a tragic situation and the duty of describing its dire outcomes to a family is painful. But imagine what it is like to deal with locked-in, to look in the eyes of someone who has a conscious mind and who is limited in her expression by the simplest of codes. The cruelty of this state is almost unrivaled in all of medicine, and neurology does offer a large list of cruel situations for us to choose from—the situation of a patient with advanced amyotrophic lateral sclerosis, also known as Lou Gehrig's disease, is no better. The solace we can take as we confront the sad reality of locked-in patients is that the profound defect of motor control reduces their emotional reactivity and seems to produce some welcome inward calm.

In terms of size, general location, and causative mechanism, locked-in syndrome is the result of damage similar to that which causes coma. But because the *precise* location of the damage is different, the result is also different, and no loss of consciousness ensues. Locked-in syndrome only occurs when the damage is located in the front part of the brain stem rather than the back (see figure 8.2). And because the pathways which bring motor signals to the entire body

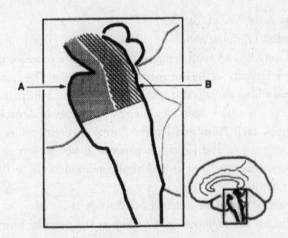

Figure 8.2. Location of brain-stem damage in cases of locked-in syndrome (A) and in cases of coma (B). The anatomical orientation is as in figure 8.1. The damage which causes the locked-in syndrome is located in the anterior (front) part of the brain stem. The damage which causes coma is located in the posterior (back) part of the brain stem.

proper are located, with only one exception, in the front part of the brain stem, the strokes that cause locked-in syndrome destroy those pathways and thus preclude any possibility of movement in virtually all muscle groups in the body. The fortunate exception concerns the pathways which control blinking and vertical eye movements because they travel separately in the back region of the brain stem. That is why they are spared in locked-in and allow some communication to take place. In short, the critical area that is damaged in coma is intact in the locked-in brain.[7]

The contrast between coma and locked-in cases offers powerful evidence for the specificity of the structures we have been considering in the generation of consciousness. But it is appropriate, at this point, to place these comments in a broader perspective of what is known about this region of the brain. In the pages ahead I suggest that explaining coma and persistent vegetative state solely in terms of damage to the ascending reticular activating system does not entirely do justice to the anatomical and functional complexity of this area.

Reflecting on the Neural Correlates of Coma and Persistent Vegetative State

We have long known with some certainty that the presence of consciousness depends on the integrity of the brain stem. The part of the brain stem whose damage disrupts consciousness and the part whose damage does not have been identified by a number of neurologists, especially by Fred Plum and Jerome Posner in their studies of comatose, vegetative, and locked-in patients. It was largely through their efforts that the latter two clinical conditions were recognized and even named.[8]

The part of the brain stem whose damage is necessary to cause coma contains the region usually known as the reticular formation. You can imagine this overall region as the eccentric axis of the tree trunk we know as the brain stem. It runs from the level of the medulla oblongata, just above the end of the spinal cord, to the top of the midbrain, just below the thalamus.[9] The part of the reticular for-

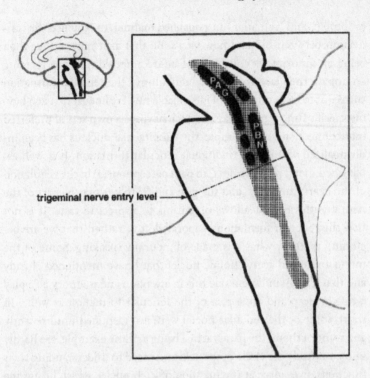

Figure 8.3. Location of some of the critical brain-stem nuclei. The anatomical orientation is the same as in figures 8.1 and 8.2. The PAG (periaqueductal gray), the PBN (parabrachial nucleus), and most of the acetylcholine and monoamine nuclei are located in the upper brain stem, within the posterior sector. This is the same overall region whose damage causes coma.

mation that is of most concern to us, however, is the part located from the level of the mid-pons on up since only from that level on up does brain-stem damage cause coma.

Some authors are reluctant to use the term "reticular formation" or "reticular nuclei" because new knowledge about the component structures reveals there is no homogeneity in the anatomy or function of the region.[10] This is precisely the same problem we face with umbrella terms such as "limbic system." On the other hand, during a transitional period, it is reasonable and helpful to refer to terms such as "limbic" and "reticular," in a qualified manner, to establish the connection between old and new views. Be that as it may, rather than being an amorphous collection of interconnected neurons forming an unpatterned lacework, i.e., a "reticulum," the reticular formation turns out to be a collection of identifiable nuclei of neurons, each having specific functions to play and each having its own sets of preferred interconnections. For example, the parabrachial nucleus has been individualized within the traditional reticular formation. It is well established that it plays a role (1) in pain perception; (2) in the regulation of the heart, the lungs, and the gut; and (3) that it may be part of the neural pathway that allows organisms to appreciate taste. It is not that the reticular formation evaporated; it is, rather, that we are beginning to know what it is made of, neurally speaking. Some of the monoamine and acetylcholine nuclei that I have mentioned already and that play an indispensable role in attention and memory also play a role in sleep and are a part of the reticular formation as well.[11] In short, some of the reticular nuclei were not identified until recently and some of them, the parabrachial being a prime example, are hardly known outside the circle of specialists devoted to understanding their functions. Eyes glaze at the mention of such nuclei, which brings me to the point in this comment: most of these newly studied nuclei that belong to the reticular formation have been identified in connection with their roles in homeostasis, which, as we previously discussed, is the regulation of the state of the internal milieu and viscera. What has mattered to the community of researchers studying them is how they contribute to, say, regulating cardiac function or intervening in reward processes or mediating pain. Their fundamental function, as far as current descriptions go in the pertinent scientific literature, is the regulation of life, the management of body states. Some of these nuclei have also been studied in relation to sleep, but most of them have not been investigated in relation to their possible role in consciousness.

What we are facing, then, is a curious split in the history of studies associated with this general region. One strand of studies, which goes back almost half a century and, regrettably, has almost been abandoned, has conceived this region as a fairly homogeneous unit and connected it with attention, arousal, sleep, and consciousness. Those studies tend to refer to the reticular formation as a unit, rather than to specific nuclei (MRF is the call acronym, the "M" standing for "midbrain" or "mesencephalic," depending on the authors, and "RF" for "reticular formation"; the acronym is not felicitous since the upper pontine reticular formation is part of the unit but is left out of the designation). A second strand of studies focuses on the function some individual nuclei play in homeostatic regulation. You might think, at first glance, that the two strands of studies are as incompatible as the respective investigators are far apart in their different specialties and laboratories. I believe, on the contrary, that the strands can be reconciled with great advantage. The different views are, in fact, delivering, unwittingly, a powerful message: the brain nuclei primarily concerned with managing the life process and representing the organism are closely contiguous to, and even interconnected with, nuclei concerned with the process of wakefulness and sleep, with emotion and attention, and ultimately with consciousness. It is even probable that some of the very same nuclei actually participate in more than one of those functions.

The Reticular Formation Then and Now

The traditional view of the reticular formation is synonymous with a collection of remarkable experiments conducted by Magoun, Moruzzi, and their colleagues in the late 1940s and early 1950s. Those experiments, in turn, were the blossoming of a pioneering tradition begun by Bremer and Jasper in the previous decade.[12]

Virtually all of these experiments were conducted on animals, mostly cats, which were under some degree of anesthesia. The typical experimental design called for (1) producing a lesion (for instance, in the preparation known as encéphale isolé, the spinal cord was separated from the brain stem by a horizontal section at the level of the medulla; in cerveau isolé, the horizontal section was placed at the junction of pons and mesencephalon); (2) stimulating a particular site electrically (for instance, a nerve or a nucleus); and (3) measuring the result of the manipulation in terms of a change in the wave patterns of the electroencephalogram. The actual behavior of the animals was not the focus of the experiments.

The upshot of these experiments was that the reticular formation was understood to constitute an activating system, which became known as the "ascending reticular activating system."

The job of the system was to maintain the cerebral cortex in an awake and alert state. This awake and alert state was then, and is now, usually taken as a synonym of consciousness. The reticular formation exerted a powerful influence on virtually all the sectors of the nervous system located above it, but especially on the cerebral cortex. The influence covered the entire span of the cerebral hemispheres, and the metaphors used to describe this influence often appealed to words such as *arousing* or *energizing*. The reticular activating system would wake up the cerebral cortex, set it in a mode of operation which would permit perception, thought, and deliberate action—in brief, make it conscious. Damage to the reticular formation would put the cerebral cortex to sleep, turn off the light on perception and thought, as it were, and preclude the execution of planned action. These metaphors are quite sensible, in general, although I do not think they tell the whole story.

The handful of contemporary scientists that have worked on the reticular formation and on its extension into the thalamus includes investigators concerned with understanding the neural basis of consciousness and attention, for example Mircea Steriade and Rodolfo Llinás, and investigators concerned with the study of sleep, for example Allan Hobson.[13] Their studies have supported the main conclusions of the Magoun and Moruzzi experiments and it can be said with certainty that the reticular formation is generally involved in sleep and wakefulness. Moreover, it is apparent that some nuclei within the reticular formation are especially involved in the generation of sleep-wakefulness cycles. This is the case, for instance, with cholinergic neurons in the pedunculopontine region and with the nuclei concerned with the distribution of norepinephrine (the locus coeruleus) and serotonin (the raphe nuclei).[14] There are intriguing details about how these different nuclei are engaged in the induction and termination of the sleep state, as well as details about their activation or silence during the particular kind of sleep in which dreams occur—rapid eye movement sleep, also known as REM sleep or paradoxical sleep. For instance, norepinephrine and serotonin neurons are silenced, but some acetylcholine neurons are very active and their activity is linked to the appearance of PGO (ponto-geniculo-occipital) waves which are found in dream sleep and are similar to the EEG waves encountered in the awake state.[15]

Recent investigations have also confirmed an important aspect of the original observations. Organisms in deep sleep produce slow and high-amplitude EEG waves, known as "synchronized" EEG, while organisms that are in awake and attentive state or in the paradoxical REM sleep state produce fast and low-amplitude waves, known as "desynchronized" EEG. But the contemporary investigators have made an important qualification to this old finding: the so-called desynchronized EEG actually hides, within it, sectors of synchronization related to small and local regions of the cerebral cortex where activity seems to be highly coordinated. In other words, as Steriade and Singer independently suggest, the term "desynchronized EEG" is a misnomer since during this state it is possible to find brain regions in which electrophysiological activity is highly synchronized.[16]

By far the most important finding confirmed by contemporary investigators is that electrical stimulation of the reticular formation causes the so-called desynchronized EEG. In other words, certain patterns of firing from the reticular formation result in the awake state or in the sleep state. The intimate connection between this region and the production of states necessary for consciousness—wakefulness and attention—is an inescapable fact. But neither the anatomical region nor the states of wakefulness and attention are sufficient to explain consciousness comprehensively.

It has also been shown that certain nuclei of the thalamus, namely the intralaminar nuclei, which happen to be the recipients of signals from the reticular formation, are an indispensable part of the pathway that produces either the awake state or the sleep state at the level of the cerebral cortex. In fact, stimulation of the MRF produces in those nuclei the same effect it causes in the cerebral cortex.[17]

Rodolfo Llinás has used this collection of findings to propose that consciousness, in both the awake state and the state of dream sleep, is generated in a closed-loop device that involves the cerebral cortex, the thalamus, and the brain-stem reticular formation. This device depends on the existence, within the reticular formation and the thalamus, of neurons that fire spontaneously. The activity of these neurons is modulated by the sensory neurons which bring signals from the outside world into the brain, but the neurons do not require signals from the outside world in order to fire in the first place. The mechanisms behind this operation are intriguing. Delivery of acetylcholine to the thalamus and the cortex changes the behavior of ion channels in the targeted neurons.[18]

In brief, the contemporary leaders of research on the reticular formation have concluded that, during conscious states, the reticular formation generates a continuous barrage of signals aimed at the thalamus and cerebral cortex, leading to the establishment of certain geometries of cortical coherence. In a parallel development, the study of the mechanisms of sleep has also shown that structures in the reticular formation are involved in the control of sleep-wakefulness cycles. Since sleep is a natural state of unconsciousness, it is reasonable to have both consciousness and sleep arise from physiological processes rooted in just about the same territory.

This is an entirely consistent set of findings, and the overall account woven around them is both coherent and valuable. This account constitutes an important advance in neuroscience, and I believe we cannot explain the neurobiology of consciousness without invoking it. But I do not believe that this is the most comprehensive account that can be proposed to relate this brain region to the phenomena of consciousness or that the neurobiology of consciousness can be fully satisfied by these findings.

Being conscious goes beyond being awake and attentive: it requires an inner sense of self in the act of knowing. Thus, the question of how consciousness emerges cannot be entirely answered by postulating a mechanism to wake up and energize the cerebral cortex, even when one specifies that, once awake, the cerebral cortex exhibits particular patterns of coherent electrophysiological activity, locally and globally. No doubt those patterns are indispensable to the conscious state. I see them as providing neural correlates for the wakeful, attentive stance during which images can be formed and manipulated and motor responses can be organized. The mere description of those electrophysiological patterns, however, does not address the issue of self and knowing, which I consider to be at the heart of consciousness. Those patterns correspond best to the tail end of the process of consciousness as I see it — the part of the process during which object maps are enhanced and the object becomes salient. Conceivably, those electrophysiological patterns may also be correlates of the processes of self and knowing. Whether this is so needs to be tested as a hypothesis aimed at specifying which part of the electrophysiological pattern would correlate with self and knowing. On the other hand, it is also possible that the aforementioned patterns (i.e., those of a globally "desynchronized" EEG, within which, under close scrutiny, local sectors of synchronization, and periodic events of nonlocal synchronization can be found) are not directly related to self and knowing, but rather to the object to be known.

My reservations over the traditional account take me to the fact I indicated at the outset of this section: there is a second strand of studies on reticular formation confronting us. In the traditional strand, reticular nuclei are involved in controlling wakefulness and attention. In the second strand, reticular nuclei, not necessarily the same ones as those targeted in the traditional studies but placed nearby and in close contact, are part of the innate machinery with which the brain regulates homeostasis and are, in order to do so, the recipients of signals that represent the state of the organism moment by moment.

A Quiet Mystery

The importance of the second strand of studies becomes apparent when we consider a mystery that has long preoccupied me: Why is it, given that the reticular formation is a long, vertically organized structure which spans the entire brain stem from the top of the spinal cord to the level of the thalamus, that only damage to a particular sector of it, from about the upper pons on upward, can cause loss of consciousness, while damage to the remainder will not alter awareness at all? This finding is well established and in no need of replication, but it has sat quietly in the literature, without much comment and without explanation. Why indeed should only one part of the reticular formation be related to the creation or suspension of consciousness, and why should that part always be the same, in case after case? And why, to project the mystery onto the experimental studies of the reticular formation, should the "ascending reticular activating system" be associated with precisely that same sector of the reticular formation? Let me try to sketch an answer.

The divider between the part of reticular formation whose damage alters consciousness and the part whose damage does not is fairly clear. You can see it well when you imagine a plane sectioning the brain stem in an orientation perpendicular to its long axis. The level for setting the plane is about the level at which the trigeminal nerve, also known as the fifth cranial nerve, happens to enter the brain stem. In their book on coma, Plum and Posner note: "The caudal extent of the structures critical to cortical arousal probably extends not much lower than the level of the trigeminal nerve entry." (See figure 8.3.)

The divider plane points to many interesting anatomical facts. First, a number of nuclei involved in high-order control of homeostasis, including the control of emotion, are located above this plane. This is true of the periaqueductal gray (PAG) nuclei, and of the parabrachial nuclei (PBN). For instance, the parabrachial nucleus (PBN), which is a recipient of signals from the entire body proper relative to the internal milieu and viscera, is located just above the divider plane, beginning at midpontine level. The nucleus pontis oralis, which receives important projections from the cerebral cortex and distributes them to this region, is also located just above the divider plane.[19] Likewise for the monoamine nuclei concerned with delivery of norepinephrine and dopamine, and for the acetylcholine nuclei. They begin to appear precisely at this level and creep upward along this region. Serotonin nuclei are also located above this area (although, unlike the nuclei for the other three transmitters, serotonin nuclei also occur at lower levels; the projections from those lower nuclei, however, are more aimed at the spinal cord than at the telencephalon).

Now, let us consider why the connection to the trigeminal nerves is possibly relevant. The trigeminal fibers carry sensory signals from the structures in the head—skin of scalp and face, muscles of both, lining of mouth and nose, in short, a comprehensive delegation from the internal milieu, viscera, and musculoskeletal aspects of the head. In sum, the trigeminal nerve contributes to the brain the last batch of information regarding the state of the organism, in a bottom to top direction, namely, the state of internal milieu, viscera, and musculoskeletal apparatus of the head.

At lower levels in the brain stem and all the way from the bottom of the spinal cord upward, segment by segment, we encounter the entry points for all the other nerves which carry signals from everywhere else in the body—limbs, chest, abdomen, everything except the head. Clearly the design for channeling signals from all over the body into the brain encompasses many entry points from the lower aspects of the spinal cord to the pons, and the totality of those signals can only reach the brain if all entry points are intact.

The anatomical clue concerns the fact that the full range of body signals conveying the current organism state is complete only *after* signals from the head enter the brain stem within the trigeminal nerve. The cranial nerves located at a higher level, respectively fourth and third, do not contribute to the integral body representation. They carry motor and autonomic commands out of the brain stem, not into. The second and first cranial nerves are, respectively, connected with vision and olfaction. They do not enter the central nervous system at brain-stem level and do not signal internal body states.

Once the signals from the trigeminal become available to a number of nuclei located both above and a bit below the entry point (the trigeminal nucleus is aligned vertically along the stem, above and below the entry point), the brain is in possession of the entire range of signals that signify body state and use a neural route, and is even in possession of some signals that signify body state and use a chemical route (they arrive via the area postrema). All that the brain is still missing, regarding current body state, are the chemical signals to be picked up by the hypothalamus and the subfornical organs. Interestingly, roughly at about this level, the brain is also in possession of auditory, vestibular, and gustatory information, and in the region above the divider level visual signals are normally available: they arrive aimed at the tectum but their subsequent projections are distributed to reticular nuclei.

This suggests that one of the powerful correlations uncovered so far between brain structure and state of consciousness relates closely to the design for the entry of body signals into the central nervous system. About and *above* the divider plane, once all neural signals and some chemical signals from the body have entered the central nervous system, a number of brain-stem nuclei concerned with regulating homeostasis have available a "comprehensive" view of the current body state, something which is vital for the regulating process. The trigeminal-nerve entry is but a clue, a pointer to the beginning of a region above which evolution would have settled to locate the life-regulation devices whose normal operation depends on data from the whole organism. My suspicion is that the classical reticular nuclei are also located above the trigeminal plane and in close proximity to life-regulation nuclei because reticular nuclei are driven by circumstances of life regulation.

When damage occurs about or above the trigeminal plane, the foundations of the proto-self are compromised, and the plotting of proto-self changes in second-order maps is also compromised. Deprived of the foundational aspects of the proto-self, the organism can no longer represent the critical substrate for knowing—the current internal state, followed by the changes that it undergoes when the organism is engaged by an object, actual or recalled. In such circumstances, independently of concomitant damage to classical reticular nuclei, the entire mechanism of consciousness should collapse. Naturally, if the classical reticular nuclei are indeed driven by proto-self structures, the compromise is compounded.

The Anatomy of the Proto-Self
in the Perspective of Classical Experiments

The results of the classical experiments on the reticular formation are compatible with the hypothesis I have been discussing regarding the neuroanatomical basis of the proto-self. In essence, four different findings must be considered. The first is the finding that in cats with encéphale isolé, which consists of dividing the brain at the junction between the spinal cord and the medulla oblongata, no changes in EEG pattern ensue. This is indeed the prediction to be drawn from my hypothesis, and it is supported by the fact that patients with damage in the medulla or in the spinal cord do not develop impairments of consciousness.

The second finding comes from the preparation known as cerveau isolé, in which the cat's brain stem is sectioned at the junction between the pons and the mesencephalon. The result is a *major* impairment: the animals are not awake, behaviorally or in terms of their EEG tracing. This is also consonant with the hypothesis and with the results of natural lesions in humans. A disruption at this level would preclude any cross signaling between the vital structures of the upper pons we have just discussed, and any other structures higher up, namely, in thalamus and cerebral cortex.[20]

The third finding is especially interesting. It concerns two types of section performed in cats at about the middle level of the pons, one immediately above the point of entry of the trigeminal nerves, the other about four millimeters higher. In the study authored by Batini, Moruzzi, and others,[21] there were two different results for the two different sections. The section just about the trigeminal level led to a permanent awake state, as indicated by the EEG, while sections just a little bit higher led to a major disturbance of wakefulness, behaviorally and in terms of EEG, no different from the consequence of severing ponto-mesencephalic sections in the cerveau isolé preparation.

Let me begin by addressing the second type of section, the one obtained about four millimeters higher than the plane of trigeminal-nerve entry. Although not as damaging as the extensive lesions that cause coma by damaging this neighborhood, such a section probably had at least three consequences: First, it damaged acetylcholine nuclei located at the level of the section and disrupted upward projections from them; second, it damaged downward cortical projections and thus precluded cortical signaling from penetrating the tegmental region of the upper pons; third, it damaged part of the parabrachial nucleus. Individually or in combination, these effects would disrupt the normal process of consciousness, for instance, by interfering with the feeding of signals to proto-self structures from both lower and higher structures. The results seen in the cat are thus consonant with the hypothesis.

Even more interesting, however, are the results of the section performed four millimeters below, at trigeminal entry level. Although we have no way of knowing what was the cats' resulting state of consciousness, their EEG profile became one of permanent wakefulness. The interpretation of this finding is as follows: First, the section precluded the sleep-producing effects of the nucleus tractus solitarius, which is located below the level of the section and is known to have hypnogenic effects; second, the section did *not* damage any of the structures which constitute the foundation of the proto-self, thus permitting signals from the cortex and thalamus to enter the critical region and alter the proto-self state. This would be possible since the animal would continue to process visual stimuli, thus activating thalamocortical and tectal regions. The visual accommodation apparatus as well as vertical eye movements would have remained intact, past memory could still be evoked from cortical structures, and all of these processes would have signaled normally to the intact brainstem region located above the section of cut. Finally, chemical information relating to overall body state would still be relayed directly to the central nervous system via the hypothalamus and the subfornical organs, and the consequences of their signaling could be brought down to the proto-self structures located above the plane of section. In short, unlike patients with coma-producing lesions, and unlike cats with sections located either slightly higher, or much higher, at the ponto-mesencephalic junction, cats with this particular section would keep intact all of the structures necessary to implement the proto-self, as well as residual means of signaling ongoing organism changes toward those structures. This situation, combined with the lack of any sleep-inducing influence from below, would explain the awake EEG, and would account for maintained wakefulness and even attention. Whether or not normal consciousness would still be possible is a question that cannot be decided on the basis of this experiment and certainly will never be answered in human beings since no natural lesion will be sufficiently circumscribed to produce such a selective defect.[22]

Reconciling Facts and Interpretations

Although they ostensibly address unrelated functions, I suspect the results of the two strands of research on the reticular formation are connected at a deep level. The two strands of studies have been motivated by different questions, but in my framework their interconnectedness begins to be visible. As an example, consider my interpretation of a recent finding in an experiment by Munk, Singer, and colleagues.[23] Munk and colleagues were able to produce in cats the sort of "desynchronized" EEG with "local synchronization" characteristics that is indicative of wakeful and attentive states. They did so by aiming their electrical stimulation at the midbrain reticular formation. However, they indicated in a footnote that they *actually stimulated the parabrachial nucleus,* something revealed by the autopsy of their experimental animals (at autopsy it is possible to follow the tracks of the stimulating electrodes and they had been placed in and about the parabrachial nucleus). In short, electrical stimulation of a nucleus of the reticular formation that has, heretofore, been associated with autonomic regulation of heart, lungs, and gut, as well as with body states such as pain, produced an electrical cortical state that is characteristic of wakefulness and attention and traditionally associated with the classical reticular nuclei.

Another experimental connection between the two strands comes from work in my laboratory in the area of emotion. In a series of studies involving healthy human subjects without neurological disease (performed in collaboration with Antoine Bechara, Thomas Grabowski, Hanna Damasio, and Josef Parvizi), we have been able to induce a variety of emotions experimentally and demonstrate, using positron emission tomography (PET), that brain-stem structures within the upper reticular formation become remarkably active with some emotions but not others.

Might this activation be a consequence of the attentive state in which the subjects need to be in order to experience these emotions? If so, our finding would be interesting but not new, given what we know from the traditional studies of the reticular formation, and given that a previous study by Per Roland and colleagues revealed activation of the reticular formation during a task requiring attention.[24] Attention alone, however, cannot explain our findings. To begin with, the control task we used demands a comparable degree of attention to imagery. If the finding we attribute to emotion had been due to attention, the activation would have vanished during the subtraction of the control task. Moreover, the findings were different for different emotions. We found maximal brain-stem activations for emotions such as sadness and anger, and little activation for an emotion like happiness. Yet, the subjects were performing the same procedure for all emotions, and there is nothing to suggest that the demand for internal attention varied across these emotions. It is likely that the upper reticular activations were tied to the neural process required for processing some specific emotions and producing the eventual feeling of those emotions.

This finding adds to the evidence suggesting that the structures of the reticular formation, traditionally linked to the control of sleep-wakefulness cycles and attention, are also linked to emotion and feeling, as well as to the representation of internal milieu and visceral states and autonomic control. There is abundant evidence that this is the case, especially regarding the periaqueductal gray (PAG). The repertoire of body changes that define several emotions is in fact controlled by PAG.[25] In brief, the structures of the so-called reticular formation of the upper pons and midbrain can be credibly linked to the notion of proto-self that I have advanced previously. That may well be the fundamental reason why they can also be associated with seemingly diverse, but nonetheless closely interrelated, functions such as emotion; attention; and, ultimately, consciousness.

Another intriguing finding from my research group comes from a study carried out in collaboration with Josef Parvizi and Gary W. Van Hoesen.[26] The study involved a detailed mapping of reticular-formation nuclei in patients with Alzheimer's disease as well as in age-matched normal controls, and it revealed a new and surprising finding: most patients with advanced Alzheimer's disease have a severe destruction of their parabrachial nucleus, on both the left and right sides of the brain stem. The parabrachial nucleus was damaged in *all* of the patients with early-onset Alzheimer's disease, an especially severe variant of the disease, and in 80 percent of the patients with late-onset Alzheimer's.

Given that patients with advanced Alzheimer's disease have a marked impairment of consciousness (see chapter 3), it is reasonable to wonder if parabrachial damage might be related to the decline in consciousness. Certainly their decline in consciousness cannot be explained by the well-known involvement of the entorhinal cortex and nearby temporal cortices.[27] Unfortunately, it is not possible at this point to go beyond wondering, because there are too many sites of focal pathology in Alzheimer's disease for one to be entirely comfortable with the correlations between particular impairments and particular sites of neural degeneration. For instance, posterior cingulate cortices and medial parietal association cortices are also heavily compromised in Alzheimer's, and they are candidate sites for second-order maps, as previously indicated.[28]

In conclusion, I see one powerful fact emerging about the critical region of the brain stem we have been discussing: it is simultaneously engaged in processes concerning wakefulness, homeostatic regulation, emotion and feeling, attention, and consciousness. The functional overlap may appear random at first glance, but upon reflection, and in the framework developed in the previous chapters, it appears sensible. Homeostatic regulation, which includes emotion, requires periods of wakefulness (for energy gathering); periods of sleep (presumably for restoration of depleted chemicals necessary for neuronal activity[29]); attention (for proper interaction with the environment); and consciousness (so that a high level of planning of responses concerned with the individual organism can eventually take place). The body-relatedness of all these functions and the anatomical intimacy of the nuclei subserving them are quite apparent.

This view is compatible with the classical idea that there is a device in the upper brain-stem region capable of creating special types of electrophysiological states in the thalamus and cortex. In fact, my proposal incorporates the classical idea but is distinctive in the following ways: first, it offers a biological rationale for the origin and anatomical placement of the device; and second, it posits that the actions of the device, as currently described, contribute importantly to the state of consciousness but do not produce the subjective aspect that defines consciousness.

ASSESSING STATEMENT NUMBER TWO: EVIDENCE FOR A ROLE OF SECOND-ORDER STRUCTURES IN CONSCIOUSNESS

Let us now turn to statement number two, which concerns damage to regions presumed to participate in the second-order neural pattern that subtends core consciousness: the cingulate gyrus, thalamic nuclei, and superior colliculi. As you read these comments recall, again, my injunctions regarding phrenology. I am not suggesting that any of these regions is solely responsible for the neural pattern that is critical for consciousness to emerge. In all likelihood, the critical neural pattern is based on cross-regional interactions.

My first choice for second-order structure is a vast portion of the cerebral cortex known as the cingulate cortex. Located near the midline, one cingulate cortex per hemisphere, this cortex is divided into numerous cytoarchitectonic regions. (See appendix figures A.4 and A.5.) In its anterior section, the structure is dominated by areas 24 and 25, immediately visible around the anterior part of the corpus callosum. Two other cytoarchitectonic areas, however, respectively areas 33 and 32, in spite of their remarkable size, are hardly visible because they are embedded in sulci. The posterior part of the cerebral cortex is constituted by area 23, quite visible on the large crown of the gyrus, and by areas 31, 29, and 30, which are again quite extensive but embedded in sulci and thus hidden.

The easiest way to summarize the known functions of the cingulate cortex is to say that they comprise an odd combination of sensory and motor roles. The cingulate is a massively somatosensory structure which receives inputs from all the divisions of the somatosensory system described in chapter 5. This includes not only a remarkable quantity of internal milieu and visceral signals but also important signals from the musculoskeletal division. Yet the cingulate is also a motor structure involved, both directly and indirectly, in the execution of a large variety of complex movements, from those that have to do with vocalization to those that involve the limbs, alone or in synergy, and to those that involve viscera. But this is not all. The cingulate is also clearly involved in the processes of attention; it is clearly involved in processes of emotion; and it is clearly involved in *consciousness*. This overlap of functions is remarkable and reminiscent of another sector of the central nervous system: the upper brain stem.

It is reasonable to say that we know both a lot and not too much about the cingulate. In spite of a number of remarkable neuroanatomical studies, the intrinsic anatomy of the cingulate and many of its connections to other regions remain uncharted territory.[30] This is also true of the neurophysiology of the cingulate, which remains some-what mysterious, especially concerning the posterior sector. One ex-planation for this panorama of ignorance has to do with the paucity of naturally occurring bilateral cingulate lesions in humans. Lesions are quite rare, as far as the anterior cingulate is concerned, and exceedingly rare in the posterior cingulate as well. Consider that not one single case has ever been described of a bilateral lesion of the cingulate in-volving *all* the cytoarchitectonic regions I enumerated above.

Under the circumstances, we should tread prudently. We know for a fact that epileptic seizures arising in the cingulate cortex are charac-terized by loss of consciousness—periods of absence that are actually longer than those caused by regular, non-cingulate seizures. A num-ber of functional neuroimaging studies have also yielded some im-portant findings. Situations in which consciousness is suspended or diminished, such as slow-wave sleep, hypnosis, and some forms of anesthesia, are associated with *reduced activity* in the cingulate cortex; on the other hand, REM sleep, as well as myriad attention paradigms are associated with *increased activity* in the cingulate cortex.[31]

In both lesion studies as well as functional imaging studies, the cin-gulate has been associated with emotion, attention, and autonomic control.[32] Bilateral anterior lesions of the cingulate cause the condi-tion known as akinetic mutism. As we saw in the case of L (chapter 3), patients with bilateral damage to the cingulate cortices have impaired consciousness although they remain awake. The patients' condition is best described as suspended animation, internally as well as externally, and this is the reason why the patients are described as akinetic and mute. From the literature and from my own observations, I can say confidently that bilateral anterior damage to the cingulate disrupts both core consciousness and extended consciousness while preserving wakefulness. We should note, however, that although the affected pa-tients do not recover an entirely normal mind, they do recover core consciousness in a matter of months. Their recovery might be due to the preservation of both posterior cingulate regions. It is possible that bilateral damage to the posterior aspect of the cingulate causes permanent damage, but I have only studied one convincing case. Be that as it may, it is reasonable to venture that bilateral damage to the entire cingulate is likely to disrupt consciousness remarkably, perhaps even permanently. Of the two large sectors of the cingulate, anterior and posterior, I would also venture that the posterior sector is the most indispensable, although I imagine that normal operations require both sectors to work in concert.

I should add that patients with damage to a region just behind and around the posterior cingulate also have disturbances of consciousness. The region is medial and parietal, a combination of retrosplenial and cuneus territories. Cytoarchitectonic areas 31, 7, and 19 are part of this region. Patients with bilateral damage to this area have a profound disturbance of consciousness. Their impairments are not as marked as those seen in coma, but are comparable to the impairments I have just described for bilateral cingulate damage.

Just as is the case with patients with bilateral cingulate damage, patients with bilateral medial parietal damage are awake in the usual sense of the term: their eyes can be open, and their muscles have proper tone; they can sit or even walk with assistance; but they will not look at you or at any object with any semblance of intention; and their eyes may stare vacantly or orient toward objects with no discernible motive. These patients cannot help themselves. They volunteer nothing about their situation and they fail to respond to virtually all the examiners' requests. Attempts to engage them in conversation are rarely successful, the results being erratic at best. We can coax them into looking briefly at an object, but the request will not engender anything else in terms of productive reaction. These patients react no differently to friends and family than they do to physicians and nurses. The notion of zombie-like behavior could perfectly well have come from the descriptions of these patients, although it did not.

The most common cause of involvement in medial parietal region is Alzheimer's disease. Outside of degenerative diseases, bilateral parietal damage is not a frequent presentation of stroke. The case of bilateral parietal damage I most vividly remember was caused by fairly symmetrical metastases from colon cancer—to picture what the patient looked like, imagine the state of absence automatism described in chapter 3 but in slow motion and without an end in sight. Head injury can cause the condition, too. The renowned British neurologist Macdonald Critchley mentioned one such case in his landmark monograph on the parietal lobes.[33]

Reflection on the anatomical specifications of the cingulate cortex indicates that it is an excellent candidate for the sort of second-order structure I proposed earlier. Its different subregions and the massiveness of its somatosensory inputs can give rise to perhaps the most "integrated" view of the entire body state of an organism at any given time. But since the cingulate cortices are also privy to signals from the main sensory channels—the appearance of an object can be reported to the cingulate easily via both thalamic projections and direct projections from higher-order cortices in inferotemporal, polar temporal, and lateral parietal regions—the cingulate could help generate a neural pattern in which the relationship between the appearance of an object and the modifications undergone by the body could be mapped in the proper causal sequence. The cingulate might actually make the critical contribution to the "feeling of knowing," the special, high-order feeling that defines core consciousness.

THE REASONS WHY the superior colliculi also qualify as a structure contributing to second-order patterns are as follows. The superior colliculi are multilayered structures which receive a multiplicity of sensory inputs from an assortment of modalities, integrate signals in a complicated fashion across their several layers, and communicate the resulting outputs to a variety of brain-stem nuclei, the thalamus, and the cerebral cortex.[34] For instance, the superior colliculi receives visual information directly from the retina in its top layer, and, just a few layers deeper, it also receives information from visual cortices; it receives auditory information from the inferior colliculi located just below, and massive somatosensory information (including visceral information) from varied brain-stem nuclei.

The integrative activity of the superior colliculi is aimed at orient-
ing the eyes, the head and neck, and the ears (in creatures that move
them) toward the source of a visual or auditory stimulus so that optimal object processing can take place. In the course of this activity, the
superior colliculi map the temporal appearance and spatial position
of an object as well as varied aspects of body state. It is conceivable that
one of their seven layers of cells might be dedicated to mapping a
second-order neural pattern describing the object-organism relationship based on the data they have available. The result would influence
classical reticular nuclei (and subsequent cortical processing, via the
intralaminar nuclei of the thalamus) as well as monoamine and
acetylcholine nuclei. In species with little cortical development this
might be the source of the simple form of core consciousness that
may accompany the execution of attentive behaviors. I hasten to add
that, in the case of humans, there is no evidence that the superior colliculi can support core consciousness in the absence of thalamic and
cingulate structures, even assuming intactness of the brain-stem
proto-self structures.[35]

FINALLY, THERE IS the matter of the thalamus. Reviewing the neuroanatomy and neurophysiology of the thalamus is outside the scope
of this book. Just as is the case for the cerebral cortex and brain stem,
the thalamus is the subject for entire books, not paragraphs. For the
sake of my argument, however, I can say that the thalamus gets first-
hand "reports" of the sequential engagement of the varied structures
representing both the characters and the events in the would-be primordial plot. The thalamus could signify the object-organism relationship in implicit form and follow that by creating more explicit
neural patterns in cingulate cortices and somatosensory cortices.
Some thalamic nuclei, such as the reticular nucleus and the pulvinar,
would be critical in this process. The idea that the thalamus is related
to consciousness is based on credible experimental evidence in animals, on the result of thalamic lesions, and on the likelihood that abnormal discharges in absence seizures, during which consciousness is

disrupted, originate in the thalamus.[36] The current evidence on the thalamus, however, is insufficient to address the hypothesis with any degree of specificity, although it is in accordance with the overall prediction. One must be content with concluding that bilateral damage to the thalamus disrupts consciousness for certain.

IN CLOSING, I will add a bit of curious and potentially relevant evidence. In the summer of 1998, my colleagues and I had a collective recognition experience when a visiting lecturer came to our department to give a talk, not about consciousness but rather about neuroimaging studies in children. In his talk the speaker included a set of images of PET scans obtained shortly after birth and within the first few months of life. Early on, the structures which are remarkably active in those newborn brains, almost as isolated islands in a sea of neuroimaging silence, are the brain stem and hypothalamus, the somatosensory cortices, and the cingulate. As you can see, the set of activated structures entirely matches those needed for the proto-self and second-order maps. The functional maturity of these structures at birth is noteworthy. Given that other brain systems have also been in full swing, e.g., auditory, the activation suggests a considerable functional precedence. The next structures to show up in PET scans, a few months later, are the ventromedial frontal lobe and the amygdala. Several of us looked at each other knowingly, and the speaker may have wondered why.[37]

ASSESSING THE OTHER STATEMENTS

Now, let us turn to the remaining statements, which concern brain sites whose damage should not cause impairment of core consciousness: the hippocampus, the higher-order cortices of temporal and frontal lobes, and the early sensory cortices of vision and hearing.

To make a long story short: bilateral damage to any of these areas individually leaves core consciousness unscathed. Sense of self and knowing still operate efficiently regarding any object that can be properly mapped. This fact underscores the following situation: proto-self and second-order maps depend largely on one set of paramidline structures—the brain stem, hypothalamic, basal forebrain, and the thalamic nuclei, as well as the centrally located cingulate cortices; while mapping of objects depends largely on less centrally located sensory cortices distributed over the cortical mantle. The left and right halves of "self and knowing" structures sit centrally, just across from each other, and are often damaged together by the same pathological cause; the left and right halves of the structures on which object mapping depends sit farther apart and are often damaged independently.

We can say with confidence that bilateral damage to the hippocampus, or to the entire anterior temporal lobe or to the entire lateral temporal lobe or to most of the medial and inferior temporal lobe does *not* cause impairments of core consciousness. HM and David, two patients we discussed in chapter 4, indicate this fact unequivocally. In fact, not even a combination of all these lesions disrupts core consciousness. Bilateral damage to the amygdalae also leaves core consciousness intact as patient S (chapter 2) shows so clearly. Needless to say, unilateral damage to any of these structures does not cause impairment of consciousness, either.

The cortege of impairments caused by all these lesions that leave consciousness intact is legend. Profound alterations of learning, memory, and language are the well-known results of such lesions. But in spite of those remarkable impairments, the patients remain keenly aware of self and surroundings, their core consciousness unscathed. They are perfectly conscious, and, more often than not, they are quite conscious of their own impairments. They are the very conscious owners of disrupted memories and broken language.

Likewise, bilateral or unilateral damage to auditory cortices, visual cortices, and prefrontal cortices does not impair core consciousness at all. In essence, the patients' ability to perceive and recognize stimuli along the auditory or visual channel is impaired, the ability to create internal images in those sensory modalities is also impaired, and there are selective memory defects pertinent to the sensory channel that has been compromised. Yet core consciousness goes on normally outside of the affected sensory modality.

Bilateral damage to early visual cortices is generally restricted to a subsector and causes visual loss either in part of the visual fields or in the entirety of the visual fields. Often it also creates one of many astonishing conditions in which visual processing is disrupted. For instance, the ability to see color may be lost across the entire visual field or in a part of it, while the ability to see movement, depth, and shape remains intact (a condition known as achromatopsia); or the ability to recognize previously familiar objects may be lost, although appreciation of the physical structure of the object remains intact (the condition known as agnosia, which we discussed previously); or the ability to survey the visual field in a harmonious and attentive manner may vanish (in what is known as Balint's syndrome).[38] In all of these instances, core consciousness remains intact; the patient is able to process normally any aspect of cognition except for the selectively disrupted aspects of visual processing. That the patients are keenly aware of what they can no longer do indicates that the "general" process of core consciousness has been spared. Of equal interest is the fact that some of these patients may retain certain aspects of nonconscious processing relative to stimuli that they can no longer either perceive or recognize. A strong example of the former occurs in the condition known as blindsight.[39] In some patients who have lost vision altogether, as a result of what is often termed cortical blindness, the patients may claim, quite truthfully, not to see any object in their visual field and yet be able, when asked to hazard a pointing finger at the possible location of the object, to move their arm and finger in the correct direction. This indicates that some correct processing is taking place such that the structures in charge of movement can guide the arm and finger in the appropriate direction even if part of the information underlying that process is not made available to the process of consciousness making.

Something along the same lines can happen in similarly blind patients when the damage to visual cortices is especially extensive, in a situation known as Anton's syndrome. The patients may deny, in the manner previously described in anosognosia, that they are blind, but the bizarre claim may have a partial explanation. The patients' eyes remain capable of verting toward objects that are attractive to a visual organism and remain capable of focusing on them. The results of the efforts of that now useless visual-perceptual machinery are of no consequence to the visual cortices themselves but are conveyed nonetheless to structures such as the superior colliculi and the parietal cortices. The brain is still informed of an ongoing set of perceptually related adjustments, probably not unlike those that would occur should the brain still be capable of visual processing.

In a situation in which visual processing is completely absent, the brain constructs a reasonably appropriate account for those perceptual adjustments that are being perceived in consciousness; an account that says, in fact, that seeing an object is in progress. The account is not adequate, of course, but is not entirely irrational, either. In the cases I have seen, such a belief generally wanes within hours, as one might expect. I am persuaded that the complete absence of visual images, actual or recalled, that occurs during the first hours of the event, explains why the patient is fooled. The profound defect in visual imagery impedes the construction of a counterargument.

I have devoted many studies, as well as *Descartes' Error,* to the situation of patients with bilateral damage to the ventromedial prefrontal lobe. I can say confidently that although their ability to decide advantageously and to resonate emotionally with certain issues is impaired, their core consciousness is not. Even bilateral damage to the dorsolateral prefrontal cortices, including the frontal pole, does not cause impairments of core consciousness.[40] Such damage does alter working memory and consequently affects extended consciousness, but these impairments leave core consciousness intact.

The "negative" evidence reported above is as important to identifying the brain territories from which consciousness can arise as the "positive" evidence concerning territories that lead to an unequivocal impairment of consciousness. Of the negative evidence just mentioned, I would like to emphasize the facts that bilateral damage to the hippocampus does not impair core consciousness, and that neither does bilateral damage to visual or auditory cortices.

The importance of the negative evidence is as follows: The hippocampus is a recipient of information from several sensory modalities and its circuitry is such that its signals can probably construct, in some fashion, an n-order map of the "scene" that results, at each moment, from the organism's multiple image-making devices. It might be conceived, then, that the hippocampus would be an ideal structure to generate the second-order map I proposed as a basis for core consciousness. This cannot be the case, however, as many studies of patients in whom the hippocampal region is damaged on both sides indicate. A profound learning and memory defect can always be found in those cases, but no impairment of core consciousness ever ensues.

CONCLUSIONS

The foregoing assessment of the available evidence allows us to draw a number of provisional conclusions.

1. Damage to the brain regions presumed to support either the proto-self or the second-order account of the organism-object relationship disrupts core consciousness. Extended consciousness is disrupted as well.
2. The regions which support either the proto-self or the second-order maps have special anatomical characteristics: (a) they are among the phylogenetically older structures of the brain; (b) they are largely located near the midline; (c) none is located on the external surface of the cerebral cortex; and (d) all are involved in some aspect of body regulation or representation.

3. Proto-self and second-order structures constitute a central re-source, and their dysfunction causes a general disruption of consciousness for any object. Early sensory structures are involved in processing separate aspects of objects, and thus the disabling of one of those structures, even if extensive, does not affect consciousness in general.

4. The regions whose damage does not cause a disruption of core consciousness constitute, in the aggregate, a larger proportion of the central nervous system than the ensemble of those that do disrupt consciousness.

5. Those same regions (e.g., early sensory cortices, higher-order cortices) are primarily involved in: (*a*) signaling the objects and the events which come to be known because of core conscious-ness; (*b*) holding records pertaining to their experience; and (*c*) manipulating those records in reasoning and creative thinking.

6. The early sensory structures are also involved in the process of making consciousness. They do so in a different manner — there is only *one set* of structures to support proto-self and second-order maps, while there are *several sets* of early sensory structures, one per sensory modality. The participation of early sensory structures includes: (*a*) initiating the process by influ-encing the proto-self structures; (*b*) signaling to second-order structures; and (*c*) being the recipients of the modulatory influ-ences consequent to the second-order neural patterns. It is be-cause of the latter influence that the enhancement of the neural patterns which support the object does occur and varied components of the object to be known become integrated.

In short, core consciousness depends most critically on the activity of a restricted number of phylogenetically old brain structures, beginning in the brain stem and ending with the somatosensory and cingu-late cortices. The interaction among the structures in this set: (1) sup-ports the creation of the proto-self; (2) engenders the second-order neural pattern which describes the relationship between the organism (proto-self) and the object; and (3) modulates the activity of object-processing regions which are not part of the set.

The specificity with which I am identifying these critical candidate sites should not be interpreted to mean that I regard any one of them as *the* basis for consciousness. None of the functions outlined above is executed at the level of a single neural site or center, but rather, these functions emerge as a result of cross-regional integrations of neural activity. I envision the sense of self and the enhancement of the object as arising out of the interactions among this set of neural sites and the set of neural sites directly involved in the construction of the object.

The neural pattern which underlies core consciousness for an object — the sense of self in the act of knowing a particular thing — is thus a large-scale neural pattern involving activity in two interrelated sets of structures: the set whose cross-regional activity generates proto-self and second-order maps, and the set whose cross-regional activity generates the representation of the object.

A Remarkable Overlap of Functions

There is a remarkable overlap of biological functions within the structures which support the proto-self and the second-order mappings. Taken individually, these structures are involved in most of the following five functions: (1) regulating homeostasis and signaling body structure and state, including the processing of signals related to pain, pleasure, and drives; (2) participating in the processes of emotion and feeling; (3) participating in processes of attention; (4) participating in the processes of wakefulness and sleep; and (5) participating in the learning process.

The entire quintet of overlaps applies fully to the brain stem and cingulate cortices, and applies in large part to the other structures. The overlaps identified here are a matter of fact, and yet they have not previously been emphasized for several reasons. Perhaps the main reason is that knowledge about one of these brain regions, the brain stem, has been segregated along two distinct strands of research, one related to the problem of homeostatic regulation and the other related to mechanisms of sleep and attention. The problems and the investigators have been kept apart. Another reason is that the neglect of emotion by neuroscience has retarded the realization that all these regions, from the brain stem to the somatosensory cortices, are critical for the processes of emotion.

It is reasonable to conclude, then, that beyond the above quintet of functions, these areas participate in one additional function: the construction of core consciousness.

The functional overlaps revealed by this survey may appear counterintuitive at first glance, and yet, after reflection on the relevant data they become transparently sensible. First, the overlaps probably result from the function of distinct "families" of contiguous nuclei. Second, notwithstanding their anatomical distinctiveness, the varied families of nuclei are highly interrelated by anatomical connections. Third, the contiguousness and anatomical interrelations which give rise to the functional overlaps are not a mere accident and probably are indicative of the overriding functional roles for the regions.

The plausibility of this idea is strengthened by considering the nature of the functional overlaps at the level of the brain stem. Regarding emotion and attention, the rationale for the functional overlap would be as follows. Emotion is critical for the appropriate direction of attention since it provides an automated signal about the organism's past experience with given objects and thus provides a basis for assigning or withholding attention relative to a given object. Simple organisms initiate wakeful behavior by having basic image-making capabilities and minimal attention, as a result of which the following happens: first, processing of objects can take place; second, emotion can ensue; third, further enhancement and focusing of attention can occur, or not occur, under the direction of emotion. In organisms capable of consciousness, the above list of events still applies, but the second step would read as follows: "Emotion can ensue and become known to the individual having it."

It makes expedient, if not necessarily tidy, housekeeping sense that structures governing attention and structures processing emotion should be in the vicinity of one another. For certain components of these processes, the structures might even be the same, although operating in slightly different modes. Moreover, it also makes good housekeeping sense that all of these structures should be in the vicinity of those which regulate and signal body state. This is because the consequences of having emotion and attention are entirely related to the fundamental business of managing life within the organism, while, on the other hand, it is not possible to manage life and maintain homeostatic balance without data on the current state of the organism's body proper.

How sensible is it for emotion and attention to overlap with core consciousness? The answer is that it is sensible, if we regard consciousness as the most sophisticated means at our disposal to regulate homeostasis and manage life. Nature is an expedient tinkerer and since consciousness is a latter-day means of achieving homeostasis, it would have been convenient for nature to evolve the machinery of consciousness *within, from,* and *in the vicinity of* the previously available machinery involved in basic homeostasis, in other words, the machinery of emotion, attention, and regulation of body states.

A New Context for Reticular Formation and Thalamus

The above conclusions do not deny, in any way, that some brain-stem structures are involved in wakefulness and attention, and that they modulate the activity of the cerebral cortex via the intralaminar thalamic nuclei, via the non-thalamic cortical projections of monoamines, and via the thalamic projections of acetylcholine nuclei. The issue is that nearby brain-stem structures and perhaps even some of the very same structures have *other* activities, namely, managing body states and representing current body states. Those activities are not incidental to the brain stem's well-established activation role: *they may be the reason why such an activation role has been maintained evolutionarily and why it is primarily operated from that region.*

In short, I have no problem with the roles that have been traditionally assigned to the brain stem's "ascending reticular activating system," and to its extension in the thalamus. On the contrary, I have no doubt that the activity of those regions contributes to creating the selective, integrated, and unified contents of the conscious mind. I simply doubt that such a contribution is sufficient to explain consciousness comprehensively. That is why I focus on a set of different, albeit related, questions: What drives those regions to perform the tasks they perform? What is the purpose of their labors? How much does the result of those labors account for what I believe consciousness is, mentally speaking?

A Counterintuitive Fact?

The above conclusions underscore an important fact: although even the simplest core consciousness requires ensemble activity that involves regions of every tier and quarter of the brain, consciousness does depend most critically on regions that are evolutionarily older, rather than more recent, and are located in the depth of the brain, rather than on its surface. In a curious way, the "second-order" processes I propose here are anchored on ancient neural structures, intimately associated with the regulation of life, rather than on the modern neural achievements of the neocortex, those which permit fine perception, language, and high reason. The apparent "more" of consciousness depends on "less," and the second-order is, in the end, a deep and low order. The light of consciousness is carefully hidden and venerably ancient.

Let me note that this is a fact, not a hypothesis—whether my hypotheses turn out to be correct or not, the fact remains that damage to these sites impairs consciousness, while damage elsewhere does not. The least that can be said about this fact is that it seems counterintuitive. We rightly think of consciousness as a significant biological advancement, even when we grant consciousness to nonhuman creatures. Well, the advancement is certainly significant, but it may be older than usually thought. What is not so old, evolutionarily speaking, is the extension of consciousness that has been allowed by memory, first, by permitting us to establish an autobiographical record; second, by giving us a broad record of other facts; and third, by endowing us with the holding power of working memory. Surely enough, these extensions of consciousness, which have blossomed so powerfully in humans, are based on the evolutionarily modern aspects of the brain, namely those of the neocortex. In the end, however, none of those astounding new features of consciousness occur independently of the modest feats of core consciousness.

PART IV

Bound to Know

Chapter Nine

Feeling Feelings

FEELING FEELINGS

I began this book by describing an obstacle: emotions cannot be known to the subject having them before there is consciousness. Now, after presenting my views on the nature of consciousness, it is time to explain how we can know an emotion. Beginning at the beginning: We know that we have an emotion when the sense of a feeling self is created in our minds. Until there is the sense of a feeling self, in both evolutionary terms as well as in a developing individual, there exist well-orchestrated responses, which constitute an emotion, and ensuing brain representations, which constitute a feeling. But we only know that we feel an emotion when we sense that emotion is sensed as happening in our organism.

The sense of "happening in the organism" comes from representing the proto-self and its changes in second-order structures. The sense of the "emotion as object" comes from representing, in structures subserving second-order representations, the activity pattern in the induction sites of emotion. Following what was outlined for other objects, I propose that: (1) the inaugural proto-self is represented at second-order level; (2) the "object" that is about to change the proto-self (the neural-activity pattern in emotion-induction sites) is represented at second-order level; (3) the ensuing changes in proto-self (enacted by "body loop" or "as if body loop" mechanisms) are also represented at second-order level.

Feeling an emotion is a simple matter. It consists of having mental images arising from the neural patterns which represent the changes in body and brain that make up an emotion. But knowing that we have that feeling, "feeling" that feeling, occurs only *after* we build the second-order representations necessary for core consciousness. As previously discussed, they are representations of the relationship between the organism and the object (which in this case is an emotion), and of the causal effect of that object on the organism.

The process that I am outlining is precisely the same we discussed for an external object, but it is difficult to envision when the object in question is an emotion, because emotion occurs within the organism, rather than outside of it. The process can only be understood when we keep in mind some of the notions introduced in the chapters on emotion (chapter 2) and on the organism (chapter 5), namely: (1) that there are several brain sites whose activity pattern induces the cortege of actions that become an emotion, and (2) that the activity pattern can be represented within second-order brain structures. Examples of emotion induction sites include nuclei in the hypothalamus, brain stem, basal forebrain, amygdala, and ventromedial prefrontal cortices. Examples of second-order structures include thalamus and cingulate cortices.

It may sound strange, at first, that feelings of emotion—which are steeped in the representation of body states, only come to be known after *other* representations of body state have been integrated to give rise to a proto-self. And it sounds strange, for certain, that the means to know a feeling is another feeling. The situation becomes understandable, however, when we realize that the proto-self, feelings of emotion, and the feelings of knowing feelings emerged at different points in evolution and to this day emerge at different stages of individual development. Proto-self precedes basic feeling and both precede the feeling of knowing that constitutes core consciousness.

THE SUBSTRATE FOR FEELINGS OF EMOTION

The collection of neural patterns which constitute the substrate of a feeling arise in two classes of biological changes: changes related to body state and changes related to cognitive state. The changes related to body state can be achieved by two mechanisms.[1] One mechanism involves what I call the "body loop." It uses both humoral signals (chemical messages conveyed via the bloodstream) and neural signals (electrochemical messages conveyed via nerve pathways). As a result of both types of signal, the body landscape is changed and is subsequently represented in somatosensory structures of the central nervous system, from the brain stem on up. The change in the representation of the body landscape can be partly achieved by another mechanism, which involves the "as if body loop." In this alternate mechanism, the representation of body-related changes is created directly in sensory body maps, under the control of other neural sites, for instance, in the prefrontal cortices. It is "as if" the body had really been changed, but it has not. The "as if body loop" mechanism bypasses the body proper, partially or entirely, and I have suggested that bypassing the body saves both time and energy, something that may be helpful in certain circumstances. The "as if" mechanisms are not only important for emotion and feeling, but also for a class of cognitive processes one might designate as "internal simulation."[2]

The changes related to cognitive state are generated when the process of emotion leads to the secretion of certain chemical substances in nuclei of the basal forebrain, hypothalamus, and brain stem, and to the subsequent delivery of those substances to several other brain regions. When these nuclei release neuromodulators in the cerebral cortex, thalamus, and basal ganglia, they cause a host of significant alterations of brain function. The most important alterations I envision include (1) the induction of particular behaviors (such as bonding and nurturing, playing and exploring); (2) a change in the ongoing processing of body states (as an example, body signals may be filtered or allowed to pass, selectively inhibited or enhanced, and their pleasant or unpleasant quality altered); and (3) a change in the mode of cognitive processing (an example of the latter, in relation to auditory or visual images, would be a change from a slow to a fast rate of image production or a change from sharply focused to vaguely focused images, a change which is an integral part of emotions as disparate as those of sadness or elation).

I suspect all three kinds of change are present in humans and in numerous nonhuman species. It is possible, however, that the third kind of change—the change in the mode of cognitive processing—is only made conscious in humans because it requires an especially high-level representation of neural events: the sort of metarepresentation of aspects of brain processing that only prefrontal cortices are likely to support.

In short, emotional states are defined by myriad changes in the body's chemical profile; by changes in the state of viscera; and by changes in the degree of contraction of varied striated muscles of the face, throat, trunk, and limbs. But they are also defined by changes in the collection of neural structures which cause those changes to occur in the first place and which also cause other significant changes in the state of several neural circuits within the brain itself.

To the simple definition of emotion as a specifically caused transient change of the organism state corresponds a simple definition for feeling an emotion: It is the representation of that transient change in organism state in terms of neural patterns and ensuing images. When those images are accompanied, one instant later, by a sense of self in the act of knowing, and when they are enhanced, they become conscious. They are, in the true sense, feelings of feelings.

There is nothing vague, elusive, or nonspecific about emotional responses, and there is nothing vague, elusive, or nonspecific about the representations which can become feelings of emotions. The substrate for emotional feelings is a very concrete set of neural patterns in maps of selected structures.

From Emotion to Conscious Feeling

In summary, the complete course of events, from emotion to feeling to feeling of feeling, may be partitioned along five steps, the first three of which were outlined in the chapter on emotion.

1. Engagement of the organism by an inducer of emotion, for instance, a particular object processed visually, resulting in visual representations of the object. The object may be made conscious or not, and may be recognized or not, because neither consciousness of the object nor recognition of the object are necessary for the continuation of the cycle.

2. Signals consequent to the processing of the image of the object activate neural sites that are preset to respond to the particular class of inducer to which the object belongs (emotion-induction sites).

3. The emotion-induction sites trigger a number of responses toward the body and toward other brain sites, and unleash the full range of body and brain responses that constitute emotion.

4. First-order neural maps in both subcortical and cortical regions represent changes in body state, regardless of whether they were achieved via "body loop," "as if body loop," or combined mechanisms. Feelings emerge.

5. The pattern of neural activity at the emotion-induction sites is mapped in second-order neural structures. The proto-self is altered because of these events. The changes in proto-self are also mapped in second-order neural structures. An account of the foregoing events, depicting a relationship between the "emotion object" (the activity at the emotion-induction sites) and the proto-self, is thus organized in second-order structures.

This perspective on emotion, feeling, and knowing is unorthodox. First, I am suggesting that there is no central feeling state before the respective emotion occurs, that expression (emotion) precedes feeling.

Second, I am suggesting that "having a feeling" is not the same as "knowing a feeling," that reflection on feeling is yet another step up. Overall, this curious situation reminds me of E. M. Forster's words: "How can I know what I think before I say it?"

The inescapable and remarkable fact about these three phenomena—emotion, feeling, consciousness—is their body relatedness. We begin with an organism made up of body proper and brain, equipped with certain forms of brain response to certain stimuli and with the ability to represent the internal states caused by reacting to stimuli and engaging repertoires of preset response. As the representations of the body grow in complexity and coordination, they come to constitute an integrated representation of the organism, a proto-self. Once that happens, it becomes possible to engender representations of the proto-self as it is affected by interactions with a given environment. It is only then that consciousness begins, and only thereafter that an organism that is responding beautifully to its environment begins to discover that *it* is responding beautifully to its environment. But all of these processes—emotion, feeling, and consciousness—depend for their execution on representations of the organism. Their shared essence is the body.

WHAT ARE FEELINGS FOR?

It might be argued that emotions without feelings would be a sufficient mechanism to regulate life and promote survival. It might be argued that signaling the results of that regulatory mechanism would hardly be necessary for survival. But that is simply not the case. Having feelings is of extraordinary value in the orchestration of survival. Emotions are useful in themselves, but the process of feeling begins to alert the organism to the problem that emotion has begun to solve. The simple process of feeling begins to give the organism *incentive* to heed the results of emoting (suffering begins with feelings, although it is enhanced by knowing, and the same can be said for joy). The availability of feeling is also the stepping stone for the next development—*the feeling of knowing that we have feelings*. In turn, knowing is the stepping stone for the process of planning specific and nonstereotyped responses which can either complement an emotion or guarantee that the immediate gains brought by emotion can be maintained over time, or both. In other words, "feeling" feelings extends the reach of emotions by facilitating the planning of novel and customized forms of adaptive response.

Now consider this: Knowing a feeling requires a knower subject. In looking for a good reason for the endurance of consciousness in evolution, one might do worse than say that consciousness endured because organisms so endowed could "feel" their feelings. I am suggesting that the mechanisms which permit consciousness may have prevailed because it was useful for organisms to know of their emotions. And as consciousness prevailed as a biological trait, it became applicable not just to the emotions but to the many stimuli which brought them into action. Eventually consciousness became applicable to the entire range of possible sensory events.

A Note on Background Feelings

What little attention has been paid to the neuroscience of emotion in the twentieth century has been concentrated on the core types of emotion studied by Darwin. Fear, anger, sadness, disgust, surprise, and happiness have been found to be universal emotions in terms of their facial expression and recognizability, as shown in the work of Ekman and others. As a result, the feelings that are most often considered are those which constitute the conscious readout of those major emotions. This would be all well and good if it would not have distracted us from the fact that we continuously have emotional feelings although those feelings are not necessarily part of the set of six "universal feelings" that hail from the six universal emotions. Most of the time we do not experience any of the six emotions, which is certainly a blessing given that four of them are unpleasant. Nor do we experience any of the so-called secondary or social emotions, a good thing, too, since they hardly fare any better in terms of pleasantness. But we do experience other kinds of emotion, sometimes low grade, sometimes quite intense, and we do sense the general physical tone of our being. I have called the readout of this background perturbation "background feelings," a term I first used in *Descartes' Error*, because these feelings are not in the foreground of our mind. Sometimes we become keenly aware of them and can attend to them specifically. Sometimes we do not and attend, instead, to other mental contents. In one way or another, however, background feelings help define our mental state and color our lives. Background feelings arise from background emotions, and these emotions, although more internally than externally directed, are observable to others in myriad ways: body postures, the speed and design of our movements, and even the tone of our voices and the prosody in our speech as we communicate thoughts that may have little to do with the background emotion. For this reason, I believe it is important to broaden our notion of the source of feelings.

Prominent background feelings include: fatigue; energy; excitement; wellness; sickness; tension; relaxation; surging; dragging; stability; instability; balance; imbalance; harmony; discord. The relation between background feelings and drives and motivations is intimate: drives express themselves directly in background emotions and we eventually become aware of their existence by means of background feelings. The relation between background feelings and moods is also close. Moods are made up of modulated and sustained background feelings as well as modulated and sustained feelings of primary emotions—sadness, in the case of depression. Finally, the relation between background feelings and consciousness is just as close: background feelings and core consciousness are so closely tied that they are not easily separable.

It is probably correct to say that background feelings are a faithful index of momentary parameters of inner organism state. The core ingredients of that index are (1) the temporal and spatial shape of the operations of the smooth musculature in blood vessels and varied organs, and of the striated muscle of heart and chest; (2) the chemical profile of the milieu close to all those muscle fibers; and (3) the presence or absence of a chemical profile signifying either a threat to the integrity of living tissues or conditions of optimal homeostasis.[3]

Thus, even a phenomenon as simple as background feelings depends on many levels of representation. For instance, some background feelings that have to do with internal milieu and viscera must depend on signals occurring as early as the substantia gelatinosa and intermediate zone of each segment of the spinal cord, and the pars caudalis of the trigeminal nerve nucleus. Other background feelings have to do with the cyclical operations of striated muscle in cardiac function and with patterns of contraction and dilation in smooth muscle which require representations in specific brain-stem nuclei such as the nucleus tractus solitarus and the parabrachial nucleus.

My notion of background feelings is similar to the notion of vitality affects presented by the developmental psychologist Daniel Stern, a notion he uses in his work with infants. That notion was first hinted at by the remarkable but unsung American philosopher Susanne Langer, a disciple of Alfred North Whitehead.[4]

THE OBLIGATE BODY-RELATEDNESS OF FEELING

Regardless of the mechanism by which emotions are induced, the body is the main stage for emotions, either directly or via its representation in somatosensory structures of the brain. But you may have heard that this idea is not correct, that in essence this was the idea proposed by William James—in brief, James proposed that during an emotion the brain causes the body to change, and that the feeling of emotion is the result of perceiving the body's change—and that time has cast the idea aside. First, there is more to my proposal than what was advanced by James. Second, the attack against James, which held sway throughout most of this century and still lingers, is just not valid, although his proposal on emotion is neither flawless nor complete.

The mechanisms I have outlined to enact emotion and produce a substrate for feelings are compatible with William James's original formulation on this theme but include many features absent in James's text. None of the features I have added undermines or violates the basic idea that feelings are largely a reflection of body-state changes, which is William James's seminal contribution to this subject. The new features I proposed add a new dimension to these phenomena, however. Even in the most typical course of events, the emotional responses target *both* body proper and brain. The brain produces major changes in neural processing that constitute a substantial part of what is perceived as feeling. The body is no longer the exclusive theater for emotions and consequently the body is not the only source for feelings, as James would have wished. Moreover, the body source may be virtual, as it were, it may be the representation of the body "as if" rather than the body "as is." I must add that I did not develop additional features or mechanisms for emotion as a means to circumvent the attacks on William James's idea, although some of my proposals do precisely that. I developed my proposals before I understood what the attackers were attacking.

One might say that there is no need to respond to the critics of William James since his seminal idea is so plausible, but that would be a mistake for several reasons. First, the account offered by William James was understandably incomplete and it must be extended in modern scientific terms. Second, part of the account that was complete was not correct in the detail. For instance, James relied exclusively on representations arising in the viscera, gave short shrift to skeletal muscles as a source for the representation of feelings, and made no mention of the internal milieu. The current evidence suggests that most feelings probably rely on all sources—skeletal and visceral changes as well as changes in internal milieu. The third reason is that the misconceptions that are part of the critique and that are still cited stand in the way of a comprehensive understanding of emotion and feeling.

Emotion and Feeling after Spinal Cord Transection

The idea that inputs from the body are not relevant to feelings is often based on the false notion that patients with spinal cord transection caused by injury should not be able to emote or feel. The problem, say the critics, is that they seem to be able to emote and to feel. Yet, only a part of the body input most relevant for feelings travels in the spinal cord. First, a considerable part of the relevant information actually travels in nerves such as the vagus, which exit and enter the brain at the level of the brain stem, well above the highest level of the spinal cord possibly damaged by an accident. Likewise, only a part of the enactment of emotions depends on the spinal cord: a large proportion of the process is mediated by cranial nerves at brain-stem level (which can act on the face and on viscera) and by other brain-stem nuclei (which can act directly on the brain above their level).

Second, a significant part of body input actually does not travel by nerves but by way of the bloodstream, again reaching the central nervous system at the level of brain stem, for instance at the area postrema, or higher.

Third, all the surveys of patients with spinal cord damage, including those that seem biased to discover an impairment of feeling and those that were biased to discover that feelings were intact, have revealed some degree of impaired feeling, as one should have expected given that the spinal cord is a *partial* conduit for relevant body input.[5] Moreover, one undisputed fact emerged in those studies: the higher the placement of damage in the spinal cord, the more impaired feeling is. This is important because the higher the section made in the spinal cord, the less input from the body will reach the brain. Higher sections should correlate with less feeling, lower sections with more. The finding would be difficult to explain were it not that some body input is, in fact, precluded by spinal cord damage. (Although it might be argued, not very credibly, that higher cord lesions by causing greater defects in movement would be accompanied by greater psychological defects and thus less feeling.)

Fourth, spinal cord transections are hardly ever complete, thus allowing for escape pathways into the central nervous system.

Fifth, some of the critics seem to conceive of the body as that part of the organism that is below the neck, the head being just forgotten. As it turns out, the face and skull, as well as the oral cavity, tongue, pharynx, and larynx—whose combination constitutes the upper portion of the respiratory and digestive tracts as well as most of the vocal system—provide a massive input into the brain. This input penetrates the brain at brain-stem level, again at a level higher than that of any spinal cord injury. Since most of the emotions express themselves prominently in changes of the facial musculature, in changes of the musculature of the throat, and in autonomic changes of the skin in the face and scalp, the representation of the related changes in the brain does not need the spinal cord for anything whatsoever and remains available as a base for feelings, even in patients with the most *complete* forms of spinal cord transection.

In conclusion, in normal circumstances we use the spinal cord both to enact a *part* of some emotions and to bring back to the brain signals about *part* of the enactment of those emotions. Accordingly, even the most complete section of the spinal cord fails to disrupt the two-way flow of signals required for emotion and feeling. The fact that any defect is found at all in spinal cord injury supports the notion that body input is relevant to the experience of emotion and feeling; such a defect can hardly be used to argue the opposite. But no one should expect Christopher Reeve not to have emotions and feelings after his accident. The fact that he has both is not evidence against the paramount role of the body in emotion and feeling.

Evidence from the Section of Vagus Nerve and Spinal Cord

The evidence from the section of the vagus nerve or of the vagus nerve and spinal cord has also been misinterpreted ever since W. Cannon turned C. S. Sherrington's experiments in dogs and his own experiments in cats into the centerpiece of his 1927 attack on James.[6] Cannon's argument is an example of the confusions that result from not distinguishing that which is external, such as an emotion, from that which is internal, such as a feeling. Why should a dog or cat, in whom the vagus nerve and spinal cord have been severed, have a complete loss of emotional display, as Cannon predicted? It should not. Severing the vagus nerve *and* the spinal cord does not impede the pathways for the responses that alter the face of the animal, such that it will display rage, fear, or peaceable cooperation with the examiner. Those responses come from the brain stem and are mediated by cranial nerves which were not compromised in Sherrington's or Cannon's experiments. Those facial expressions remained intact after combined sections of the vagus and of the spinal cord, as they should. Dogs responded angrily when shown cats and vice versa, even if they could not move their bodies, which were paralyzed below the neck. (Incidentally, if those animals had been stimulated electrically in the appropriate brain sites, they would have shown the phenomenon known as "sham rage," a display of unmotivated expressions of anger.)

But what about the animal's feelings? They certainly could not be tested, but based on the ideas I have proposed, those feelings were probably altered in part—the animals would receive signals from their facial expressions and would have intact signaling from brain-stem nuclei, both of which would be a base for feeling, but they would not receive visceral input which would have been based on signals from the vagus nerve and the spinal cord. At this point, Cannon threw caution to the winds and wondered if feelings could possibly be far when there was so much of an emotional display. He took the presence of emotion as a sure sign for the presence of feeling. The error rests entirely with the failure of making a principled distinction between emotion and feeling and of recognizing the sequential, unidirectional enchainment of the process—from inducer, to automated emotion, to representation of emotional changes, to feeling.

Lessons from Locked-In Syndrome

One of the most intriguing, albeit indirect, lines of evidence for the importance of body input in the generation of feelings comes from locked-in syndrome. As discussed in chapter 8, locked-in occurs when a part of the brain stem such as the pons or midbrain is damaged anteriorly, in its ventral aspect, rather than posteriorly, in its dorsal aspect. The motor pathways which convey signals to the skeletal muscles are destroyed, and only one pathway for vertical movement of the eyes is spared, sometimes not completely. The lesions that cause locked-in are placed directly in front of the area whose lesions cause coma or persistent vegetative state, yet locked-in patients have an intact consciousness. They cannot move any muscle in their face, limbs, or trunk, and their communication ability is usually limited to vertical movements of the eyes, sometimes one eye only. But they remain awake, alert, and conscious of their mental activity. The voluntary blinking of these patients is their sole means of communicating with the outside world. Using a blink to signify a letter of the alphabet is the laborious technique with which locked-in patients compose words, sentences, and even books, slowly dictated—one should say blinked—to an attentive note taker.

A remarkable aspect of this tragic condition and one that has been neglected to date is that although patients are plunged, fully conscious, from a state of human freedom to one of nearly complete mechanical imprisonment, they do not experience the anguish and turmoil that their horrifying situation would lead observers to expect. They have a considerable range of feelings, from sadness to, yes, joy. And yet, from accounts now published in book form, the patients may even experience a strange tranquillity that is new to their lives. They are fully aware of the tragedy of their situation, and they can report an intellectual sense of sadness or frustration with their virtual imprisonment. But they do not report the terror that one imagines would arise in their horrible circumstances. They do not seem to have anything like the acute fear experienced by so many perfectly healthy and mobile individuals inside a magnetic resonance scanner, not to mention a crowded elevator.[7]

My way of explaining this surprising finding is as follows: Blinking and vertical eye movements aside, the damage in locked-in precludes any motion, either voluntary or enacted by emotional responses, of any part of the body. Facial expression and bodily gestures in response to a deliberate intention or an emotion are precluded (there is only a partial exception—tears can be produced although the motor accompaniments of crying are missing). Under the circumstances, any mental process which would normally induce an emotion fails to do so through the "body loop" mechanism we have discussed. The brain is deprived of the body as a theater for emotional realization. Nonetheless, the brain can still activate emotion-induction sites in the basal forebrain, hypothalamus, and brain stem, and generate some of the internal brain changes on which feelings depend. Moreover, since most signaling systems from body to brain are free and clear, the brain can get direct neural and chemical signaling from organism profiles that fit background emotions. Those profiles are related to basic regulatory aspects of the internal milieu and are largely uncoupled from the patient's mental state because of brain-stem damage (only the bloodstream chemical routes remain open both ways). I suspect that some of the internal-milieu states are perceived as calm and harmonious. Support for this idea comes from the fact that when these patients have a condition which ought to produce pain or discomfort, they can still register the presence of that condition. For instance, they feel stiff and cramped when they are not moved by others for a long time. Curiously, the suffering that usually follows pain seems to be blunted, perhaps because suffering is caused by emotion, and emotion can no longer be produced in the body theater: it is restricted to "as if body" mechanisms.

Another line of evidence corroborating this interpretation comes from patients undergoing surgery who receive an injection of curare, a substance that blocks the activity of skeletal muscles by acting on the nicotinic receptors of acetylcholine. If curare acts before the proper induction of anesthesia suspends consciousness, the patients become aware of their paralysis. Like patients with locked-in, curarized patients are able to hear the conversations of those around. Based on reports obtained after the event, these patients are less calm than patients with locked-in and closer to what one might expect if one imagines being in the same situation. There may be a clue to explain the difference. Curare blocks the nicotinic receptors of acetylcholine, the transmitter that is necessary for nerve impulses to contract muscular fibers. Since the skeletal muscles throughout our face, limbs, and trunk are of the striated type and have such nicotinic receptors, curare blocks neurochemical impulses at the site of all those neuromuscular junctions and causes paralysis. However, the nerve impulses that lead smooth muscles to respond under the autonomic control of emotions use muscarinic receptors that are *not* blocked by curare. Under the circumstances, it is possible for one part of the emotional responses, that which depends on pure autonomic signals, to be enacted in the body theater and be represented back in neural structures.

As a whole, this evidence suggests that the "body loop" mechanism of emotion and feeling is of greater importance for real experience of feelings than the "as if body loop" mechanism that I have proposed as an alternate and complement.

Learning from Emotion with the Help of the Body

A recent series of learning experiments also provides evidence for the role of the body in emotion. It has been demonstrated, in both rats and humans, that recall of new facts is enhanced by the presence of certain degrees of emotion during learning. James McGaugh and his colleagues have led these studies whose results are now well confirmed.[8] For instance, if you are told two stories of comparable length that have a comparable number of facts, differing only because in one of them the facts have a high emotional content, you will remember far more detail from the emotional story than from the other. You may be pleased to know that we have this in common with rats when they are placed in an equivalent situation. They, too, have better success in a standard learning situation when a certain amount of emotion happens at the right time. Now, after the vagus nerves of the rats are severed, emotion no longer helps their performance. Why so? Well, without the vagus, the rats are also deprived of substantial visceral input to the brain. It must be the case that the particular visceral input now missing is vital for the sort of emotion that assists learning.

Chapter Ten

Using Consciousness

Unconsciousness and Its Limits

There is a growing agreement among those who think about the problem of consciousness that consciousness is valuable and that it prevailed in evolution because of that value. There is less agreement, however, when it comes to the precise contribution that consciousness has made.

I began this book by calling attention to the unconscious nature of the emotions and showing how efficacious emotions and feelings can be, even when organisms do not know of their existence. It is reasonable to ask, then, what possible advantage can organisms derive from knowing that those emotions and feelings are taking place? Why is consciousness beneficial? Might we have been equally successful as living creatures without knowing that we have feelings?

I began addressing these questions in the previous chapter but a more detailed answer requires a consideration of the powers and limits of unconscious processing. I do not need to argue that both the thoughts currently present in our minds and the behaviors we exhibit are the result of a vast amount of processing of which we are not aware. The influence of unknown factors on the human mind has long been recognized. In antiquity, the unknown factors were called gods and destiny. Earlier in this century, the unknown factors came closer to our beings and were located in the subterranean of the mind. In the version usually identified with Sigmund Freud, a certain set of early individual experiences would have shaped the working of the subterranean. In another version, Carl Jung's, the shaping of the sub-terranean would have begun long ago in evolution. We do not need to endorse the mechanisms proposed by either Freud or Jung to ac-knowledge the existence and recognize the power of unconscious processes in human behavior. Throughout the century, and through work unrelated to the original proposals of Freud and Jung, the evidence for unconscious processing has not ceased to accumulate.

The field of social psychology has produced massive evidence for nonconscious influences in the human mind and behavior. The telling examples are too numerous to list but comprehensive reviews by J. Kihlstrom and A. Reber provide a good entry into the fascinating facts.[1]

Cognitive psychology and linguistics have produced their own powerful evidence.[2] For example, by the age of three, children make amazing usage of the rules of construction of their language, but they are not aware of this "knowledge," and neither are their parents. A good example comes from the manner in which three-year-olds form the following plurals perfectly:

$$\text{dog} + \text{plural} = \text{dog } z$$
$$\text{cat} + \text{plural} = \text{cat } s$$
$$\text{bee} + \text{plural} = \text{bee } z$$

The children add the voiced z, or the voiceless s, at the end of the right word but the selection does not depend on a conscious survey of that knowledge. The selection is unconscious. The knowledge of gram-matical structure, to which Noam Chomsky's work pointed us in midcentury, is not consciously present in most instances of its perfectly correct and effective usage.[3]

The examples from the field of neuropsychology are equally numerous and telling. For instance, the knowledge acquired through conditioning remains outside conscious survey and is expressed only indirectly; patients who can no longer consciously recognize faces can detect familiar faces nonconsciously; legally blind patients with certain brain lesions are able to point relatively accurately to a source of light that they cannot consciously see.[4] The retrieval of sensorimotor skills without consciousness of the knowledge expressed in the movement provides a good illustration of this situation.

The term *sensorimotor skill* refers to the sort of thing you acquire when you learn to swim, ride a bike, dance, or play a musical instrument. The learning of such skills involves multiple executions during which the performance of the task is progressively perfected. You do not learn to play the violin with one lesson, even if you happen to be the new Heifetz. It requires multiple trials. On the other hand, you can learn my face and my name in one shot.

There are reliable tasks to measure skill learning in the laboratory, such as mirror tracing or rotor pursuit. In the latter, for instance, you are asked to hold the tip of a stylus in contact with a minute dot, marked at the edge of a circular plate, while the plate keeps gyrating at fast speed. It takes time and several trials to master a good performance, which consists of keeping precise pace with the circular motion of the plate. It requires a fine coordination between the speed of the plate and the speed of arm movement. A computer automatically measures the performance by sensing the amount of time the stylus is in actual contact with the small dot.

Healthy individuals master this task in just a few sessions and when we plot the measurements of the performances across those sessions, we realize that there is a learning curve. The next session always has fewer errors than the session before, and the time needed to complete the task gets shorter. Normal subjects are thus learning a number of things concurrently. They are learning about the place and the people who are administering the experiment; they are learning about the apparatus for the experiment; they are learning the instructions for the task; and they are learning to perform the task better and better. Practice does, indeed, make perfect, as mother always said, and eventually one cannot get any better: practice can get you to Carnegie Hall.

Now, let us repeat the experiment but change the participants, specifically patients with severe amnesia, such as David, who cannot learn any new face, or place, or word, or situation. You might expect that those patients would be unable to learn the task, but that is not so. They learn it perfectly and their actual performance is in no way distinguishable from the performance of the normal subjects. There is, however, a major difference between David, on the one hand, and the normal subjects: it pertains to what surrounds the performance rather than to the performance itself. The amnesic patients do not learn anything whatsoever about the place, the people, the apparatus, and the instructions for the experiment. All that they learn is to perform the task, and they need to be told, ever so gently, every time they confront the apparatus, what the task is all about. That they do it, and do it better and better each time, with fewer errors and at faster speed, is a clear indication that the deployment of the skill does not depend on the conscious survey of the facts describing the task. David does not remember what he thought about the difficulties he encountered in the first sessions, nor does he remember what he thought about how to correct the performance and hone the skill. He simply performs in a skilled manner. For him, as a conscious person, it is as if the situation is being encountered for the first time. And yet, outside of conscious survey of both instructions and skill knowledge, his brain is ready to deploy that skill.

No less remarkable is a fact that we were also able to demonstrate in these patients: knowledge of the skill remains available long after it was acquired. For instance, David could still perform as well as normal controls two years after skill acquisition. This indicates that knowledge had been consolidated.

You might say that while nonconscious skill execution such as this is interesting, it is of no worth to the patients and irrelevant to normal individuals. After all, we usually know the circumstances in which we learn a skill and the events connected with the learning. But the fact that sensorimotor skills can be deployed with little or no conscious survey is of great advantage in the performance of numerous tasks, minor and not so minor, in our daily lives. The lack of dependence on conscious survey automates a substantial part of our behavior and frees us in terms of attention and time—two scarce commodities in our lives—to plan and execute other tasks and create solutions for new problems.

Automation is also of great value in expert motor performances. Part of the technique of a fine musician or athlete can remain underneath consciousness, allowing the performer to concentrate on the higher-level guidance and control of the technique so as to perform according to the particular intention formulated for a certain piece.

WHEN A FACE-AGNOSIC patient (such as Emily, the patient I discussed in chapter 5) is shown, in random presentation, faces of people whom she has never met as well as faces of close relatives and friends, and when we simultaneously record her skin conductance with a polygraph, a dramatic dissociation takes place. To her conscious mind, the faces are all equally unrecognizable. Friends, relatives, and the truly unfamiliar generate the same void, and nothing comes to mind to permit the discovery of their identity. And yet, the presentation of virtually every face of a friend or relative generates a distinct skin-conductance response, while unknown faces do not. None of these responses is noticed by the patient. Moreover, the magnitude of the skin-conductance response is higher for the closest of relatives.

The interpretation is unequivocal. In spite of being unable to conjure up knowledge in image form, such that conscious survey would permit recognition, the patient's brain can still produce a specific response that occurs outside of conscious survey and betrays past knowledge of that particular stimulus. The finding illustrates the power of nonconscious processing, the fact that there can be specificity underneath consciousness.

PERHAPS THE MOST decisive example of high-level nonconscious processing comes from work performed in my laboratory in collaboration with Antoine Bechara and Hanna Damasio. The work requires a decision-making task and reveals that a number of decisions that can eventually be reached by using relevant knowledge and logic are facilitated by a nonconscious influence prior to knowledge and logic playing their full roles. It also reveals that emotions play an important role in driving the nonconscious signals. The task involves a game of cards, in which, unbeknownst to the player, some decks are good and some decks are bad. The knowledge as to which decks are good and which are bad is acquired gradually, as the player removes card after card from varied decks. The source of the knowledge is the fact that the picking of certain cards from certain decks leads to financial rewards or penalties. We began using this task to investigate decision making in patients with frontal lobe damage and recently we have used it to investigate emotion and consciousness both in patients with brain damage and in healthy individuals without neurological disease.

By the time normal players begin choosing consistently the good decks and begin avoiding the bad decks, they have no conscious depiction of the situation they are facing and have not formulated a conscious strategy for how to deal with the situation. At that point, however, the brains of these players are already producing systematic skin-conductance responses, immediately prior to selecting a card from the bad decks. No such responses ever appear prior to selecting cards from the good decks. These responses are indicative of a nonconscious bias, obviously connected with the relative badness or goodness of the decks. How the brain "gets to know," without consciousness, that some decks are good and some decks are bad is the critical question. In the narrow sense of knowing, the brain does know the following implied associations: things that are rewarding cause pleasant states; things that are punishing cause unpleasant states; thus a certain object that is a consistent source of punishment is to be avoided. In this arrangement, the facts of past experience do not need to be made conscious. They do need to be connected by appropriate neural patterns with the current situation so that their preset influence can be exerted as a covert bias.[5] Yet, conscious humans can go beyond the state of processing described above. Not only can humans become conscious of the biases, i.e., know, in the broad sense, they can also reach appropriate conclusions through conscious reasoning and use those conclusions to avoid unpleasant decisions.

We know from the situation of patients who lose the covert biasing system—patients with damage to the ventromedial prefrontal cortex or to the amygdala—that the decision apparatus is impoverished to a dramatic degree. This indicates that the nonconscious system is deeply interwoven with the conscious reasoning system such that the disruption of the former leads to an impairment of the latter. But in the situation of a person without neurological disease, in which both the nonconscious and conscious systems are present and normal, it is apparent that the conscious component extends the reach and efficacy of the nonconscious system. Consciousness allows the player to discover if the strategy is correct and, in case it is not, to correct the strategy. Moreover, consciousness allows the player to represent the context of the game and decide if he or she should stop playing it or wonder about the possible value of the situation for the player or for the examiner.

THE MERITS OF CONSCIOUSNESS

What is consciousness really good for, considering that so much adequate regulation of life can be achieved without conscious processing, that skills can be automated and preferences enacted without the influence of a knowing self? The simplest answer: consciousness is good for extending the mind's reach and, in so doing, improving the life of the organism whose mind has that higher reach.

Consciousness is valuable because it introduces a new means of achieving homeostasis. I am not referring to a more efficient means of balancing the internal milieu than the entirely nonconscious machinery we have long had in place in the brain stem and hypothalamus. Rather, I am referring to a new means of solving different kinds of problems that are connected, nonetheless, to the problems solved by previously existing means of homeostatic regulation. In other words, devices in the brain stem and hypothalamus can coordinate, nonconsciously and with great efficiency, the jobs of the heart, lungs, kidneys, endocrine system, and immunological system such that the parameters that permit life are maintained within the adequate range, while the devices of consciousness handle the problem of how an individual organism may cope with environmental challenges not predicted in its basic design such that the conditions fundamental for survival can still be met.

A fact compatible with this conclusion is the mismatch between the demands of the environment and the degree to which organisms can cope with these demands by means of automated and stereotyped devices. Nonconscious creatures are capable of regulating homeostasis internally and equally capable of breathing the air and finding the water and transforming the energy required for survival within the sort of environment to which they are suitably matched by evolution. Creatures with consciousness have some advantages over those that do not have consciousness. They can establish a link between the world of automatic regulation (the world of basic homeostasis that is interwoven with the proto-self) and the world of imagination (the world in which images of different modalities can be combined to produce novel images of situations that have not yet happened). The world of imaginary creations—the world of planning, the world of formulation of scenarios and prediction of outcomes—is linked to the world of the proto-self. The sense of self links forethought, on the one hand, to preexisting automation, on the other.

Consciousness is not the sole means of generating adequate responses to an environment and thus achieving homeostasis. Consciousness is just the latest and most sophisticated means of doing so, and it performs its function by making way for the creation of novel responses in the sort of environment which an organism has not been designed to match, in terms of automated responses.

I would say that consciousness, as currently designed, constrains the world of imagination to be first and foremost about the individual, about an individual organism, about the self in the broad sense of the term. I would say that the effectiveness of consciousness comes from its unabashed connection to the nonconscious proto-self. This is the connection that guarantees that proper attention is paid to the matters of individual life by creating a *concern*. Perhaps the secret behind the efficacy of consciousness is selfness. In short, the power of consciousness comes from the effective connection it establishes between the biological machinery of individual life regulation and the biological machinery of thought. That connection is the basis for the creation of an individual concern which permeates all aspects of thought processing, focuses all problem-solving activities, and inspires the ensuing solutions. Consciousness is valuable because it centers knowledge on the life of an individual organism.

Evidence for the value of consciousness comes from considering the results of even its mildest impairments. When the mental aspect of self is suspended, the advantages of consciousness soon disappear. Individual life regulation is no longer possible in a complex environment. In the full personal and social sense, individuals remain capable of basic and immediate bodily maintenance. But their connection to the environment on which they depend is broken down, and, because of the breakdown, they cannot sustain such bodily maintenance. In fact, left to their own devices, death would ensue in a matter of hours because bodily maintenance would collapse. This, and comparable examples, suggest that a state of consciousness which encompasses a sense of self as conceptualized in this book is indispensable for survival.

The imagetic level of "self in the act of knowing" is advantageous for the organism because it orients the entire apparatus of behavior and cognition toward self-preservation, as Spinoza would have wished, and eventually toward cooperation with the other, as we must wish.

WILL WE EVER EXPERIENCE THE CONSCIOUSNESS OF ANOTHER?

I am often asked if, as a consequence of our greater understanding of consciousness, we will eventually be able to gain access to each other's mental experiences. My answer to the question has long been no, and my opinion has not changed. This may sound surprising at first glance, given that we are gathering so many new facts about neurobiology. However, as I see it, no amount of knowledge about the biology behind mental images is likely to produce, in the mind of the possessor of the knowledge, the equivalent of the experience of any mental image in the mind of the organism that creates it.

Imagine that, in a future that may not be too distant, an amazing new scanner allows you to scan my brain in unprecedented depth as I look, say, at San Francisco Bay. There we are, you, me, the amazing scanner, and San Francisco Bay. The scanner will focus not just on the level that is currently available, that of the so-called large-scale systems, but at a far deeper level. Imagine, for instance, that you can scan my retinas, my lateral geniculate nuclei, and all of the early visual cortical regions, separately and at different times, during the buildup of the visual image I am now forming of the sight before me. Furthermore, imagine that the scanning can take you to different cell layers of the varied cerebral cortices and subcortical nuclei, and that the spatial resolution is so good that you can see with clarity the patterns of neuron firings that correspond to the things both you and I can look at outside our organisms. Imagine, finally, to push this science-fiction scenario beyond the current envelope but by no means beyond the plausible envelope, that your amazing scanner also provides you with a description of the physics and chemistry of the neural-activation patterns that you detect in my varied neuron ensembles.

Armed with the data from all of these high-powered scans and assuming that you have the equally high-powered computers to analyze the wealth of data in some meaningful way, you may well obtain a remarkable set of *correlates* of the contents of the image in mind. I am submitting to you, however, that by no means will you have obtained my *experience* of that image. This is a key issue to clarify in any discussion of the neurobiology of consciousness and mind. You and I can have an experience of the same landscape, but each of us will generate that experience according to our own individual perspective. Each of us will have a separate sense of individual ownership and individual agency. As you look at the patterns of activity in my brain which underlie my experience of San Francisco Bay, you are having your own personal experience of all those neural data but not my experience of San Francisco Bay. You have an experience of something that is highly correlated with my experience, but it is an experience of something different. You do *not* see what I see when you look at *my* brain activity. *You see a part of the activity of my brain as I see what I see.*

My own experience of the landscape comes easily, cheaply, and directly, with no need of intervening technology. I do not need to know a thing about the particular behavior of neurons and molecules in different areas of my brain in order to have the experience of San Francisco Bay. In fact, even when I recall in my mind all the knowledge of neurophysiology that I have pertinent to forming mental visual images of landscapes, it does not make one bit of difference to the forming of these current images or to my experience of them. It is nice to know a little bit about how the brain does its job, but it is not necessary at all to experience anything. It will be even nicer to know more about the brain but not because that will be helpful at all to experience the world.

The point should be clear then: We will know more and more about the physiology of mental image processing and that will give us a better and better understanding of the mechanisms behind mind and consciousness. That is perfectly compatible with the fact that such knowledge is not necessary for the experience of any images.

Now comes another problem. The fact that knowledge of the biology of image processing is irrelevant for the experience of those images is often taken to mean that it is simply not possible to discover the biology behind those images. Of course, the former claim has nothing to do with the latter. We have seen that our knowledge of the biological mechanisms behind the formation of images and their experience is one thing and our experience of those images is another. As far as we can fathom, no amount of knowledge about the neurophysiology of the formation and experience of mental images will ever produce the experience of those mental images in those who possess that knowledge, although greater knowledge will give us a more satisfactory explanation of how we come to have such experiences of images.

The philosopher Frank Jackson introduced a story about this problem that has become quite well known in philosophical circles and is often cited in discussions on this issue.[6] The story tells of Mary, a card-carrying neuroscientist, who has grown up in an enclosed black-and-white environment without ever experiencing colors, although she happens to know every available fact about the neurophysiology of color vision. One day Mary leaves her colorless cocoon, comes out into the real world, and experiences color for the first time, an entirely new and surprising thing for her. The first traditional point of this story is that Mary's superior knowledge of the neurophysiology of color had never given her the experience of color. So far so good. Unsurprisingly, I agree that such should be the case, according to what I explained above. Now, for the second and main point of the story, the one with which I cannot agree: the fact that Mary had never experienced color in spite of all her abundant knowledge of its biological underpinnings is taken as meaning that neurophysiological knowledge cannot be used to explain mental experience, that there is an abyss between knowledge and experience that cannot be bridged scientifically.

I disagree with these conclusions on several counts. The first and most important is that explaining the mechanisms behind an experience and having the experience are entirely different matters as the little fiction with which I began this section illustrates. We should not conclude that neurophysiological knowledge is inadequate to explain the phenomenon just because having that neurophysiological knowledge is not equal to the experience of the phenomenon we are trying to explain. It should not be and could not be. The second reason for disagreement follows from the arguments presented earlier. The experience of a particular stimulus, including color, depends not just on the formation of an image but also on the sense of self in the act of knowing. Mary's fable is inadequate for the purpose it is used because it never deals neurophysiologically with the matter of her experience of color but simply with her formation of an image of color.[7]

Now, Mary could, of course, become knowledgeable about the neural underpinnings of consciousness. She might read this book. At that point, she would know something about how to explain general mechanisms of the mental experience of color, but that would still not allow her to have an experience of color. *Explaining* how to make something mental or something ours in scientific terms is an entirely different matter from *making* that something mental and ours *directly*.

THE RESISTANCE FOUND in some scientific quarters to the use of subjective observations is a revisitation of an old argument between behaviorists, who believed that only behaviors, not mental experiences, could be studied objectively, and cognitivists, who believed that studying only behavior did not do justice to human complexity. The mind and its consciousness are first and foremost private phenomena, much as they offer many public signs of their existence to the interested observer. The conscious mind and its constituent properties are real entities, not illusions, and they must be investigated as the personal, private, subjective experiences that they are.

The idea that subjective experiences are not scientifically accessible is nonsense. Subjective entities require, as do objective ones, that enough observers undertake rigorous observations according to the same experimental design; and they require that those observations be checked for consistency across observers and that they yield some form of measurement. Moreover, knowledge gathered from subjective observations, e.g., introspective insights, can inspire objective experiments, and, no less importantly, subjective experiences can be explained in terms of the available scientific knowledge. The idea that the nature of subjective experiences can be grasped effectively by the study of their behavioral correlates is wrong. Although both mind and behavior are biological phenomena, mind is mind and behavior is behavior. Mind and behavior can be correlated, and the correlation will become closer as science progresses, but in their respective specifications, mind and behavior are different. This is why, in all likelihood, I will never know your thoughts unless you tell me, and you will never know mine until I tell you.

WHERE DOES CONSCIOUSNESS RANK IN THE GRAND SCHEME?

The conflation of so many meanings around the word *consciousness* renders it almost unusable without qualification, and this conflation is probably responsible for the supreme status to which consciousness has been elevated. The conflation has led to the unrestrained attribution to consciousness of properties of the human mind that we consider extremely refined and uniquely human, such as our ability to distinguish good from evil, our knowledge of the needs and wants of fellow humans, our sense of the place we occupy in the universe. The attribution has rendered consciousness untouchable. I see consciousness, instead, as allowing the mind to develop the properties we so admire but not as the substance of those properties. Consciousness is *not* conscience. It is *not* the same as love and honor and mercy; generosity and altruism; poetry and science; mathematical and technical invention. Nor, for that matter, are moral turpitude, existential angst, or lack of creativity examples of bad states of consciousness. The consciousness of most criminals is not impaired. Their conscience may be.

The marvelous achievements that come from the human mind require consciousness in the same fundamental way that they require life, and that life requires digestion and a balanced internal chemical milieu. But none of those marvelous achievements is directly caused by consciousness. They are, instead, a direct consequence of a nervous system which, being capable of consciousness, is also equipped with a vast memory, with the powerful ability to categorize items in memory, with the novel ability to code the entire spectrum of knowledge

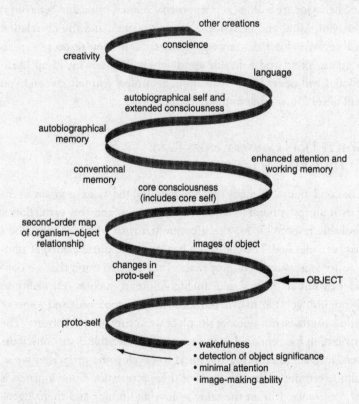

Figure 10.1. From wakefulness to conscience

in language form, and with an enhanced ability to hold knowledge in mental display and manipulate it intelligently. Each of these abilities, in turn, can be traced to myriad mental and neural components.

Core consciousness does not rank especially high in the order of operations which permit human beings to be what they are. It is part of the foundation of a complicated edifice, not one of the dreamy spires at its top. In rank order, core consciousness sits above, but not far from, other foundational capacities, such as action, emotion, and sensory representation, which we share with several nonhuman species.

The essence of those foundational capacities has probably changed little when we compare the human version to the nonhuman. For example, I see no evidence that emotion has become "better" in humans. What has become different is our sense of the role emotions play in our lives, and that difference is a consequence of the greater knowledge we have of the substance of our lives. Memory, language, and intelligence make the difference, not emotion. The same probably applies to consciousness. Extended consciousness occurs in minds endowed with core consciousness, but only when those minds can rely on superior memory, language, and intelligence, and when the organisms which construct those minds interact with suitable social environments. In short, consciousness is a grand permit into civilization but not civilization itself.

When I bring consciousness down from its current pedestal, I am not bringing the human mind down from its pedestal. It is just that what put the human mind on its pedestal and should keep it there are not only the biological phenomena subsumed by the term consciousness, but also many other phenomena which we need to describe, name, and attempt to understand scientifically. Nonetheless, I am ready to admit that we probably were banished from Eden because of consciousness. Consciousness is not the full taste of the fruit of knowledge, but innocent consciousness did start things along the way, many species ago and many millions of years before humans began to construct conceptions of their own nature.

Chapter Eleven

Under the Light

By Feeling and by Light

Perhaps the most startling idea in this book is that, in the end, consciousness begins as a feeling, a special kind of feeling, to be sure, but a feeling nonetheless. I still remember why I began thinking of consciousness as feeling and it still seems like a sensible reason: consciousness *feels* like a feeling, and if it feels like a feeling, it may well be a feeling. It certainly does not feel like a clear image in any of the externally directed sensory modalities. It is not a visual pattern or an auditory pattern; it is not an olfactory or gustatory pattern. We do not see consciousness or hear consciousness. Consciousness does not smell or taste. Consciousness feels like some kind of pattern built with the nonverbal signs of body states. It is for this reason perhaps that the mysterious source of our mental first-person perspective—core consciousness and its simple sense of self—is revealed to the organism in a form that is both powerful and elusive, unmistakable and vague.

The seventeenth-century French philosopher Malebranche might have approved of this account since he wrote as follows three hundred years ago:

> It is through light and through a clear idea that the mind sees the essence of things, numbers, and extensions. It is through a vague idea or through feeling that the mind judges the existence of creatures and that it knows its own existence.[1]

The idea of consciousness as a feeling of knowing is consistent with the important fact I adduced regarding the brain structures most closely related to consciousness: such structures, from those that support the proto-self to those that support second-order mappings, process body signals of one sort or another, from those in the internal milieu to those in the musculoskeletal frame. All of those structures operate with the nonverbal vocabulary of feelings. It is thus plausible that the neural patterns which arise from activity in those structures are the basis for the sort of mental images we call feelings. The secret of making consciousness may well be this: that the plotting of a relationship between any object and the organism becomes the feeling of a feeling. The mysterious first-person perspective of consciousness consists of newly-minted knowledge, information if you will, expressed as feeling.

Presenting the roots of consciousness as feelings allows one to glean an explanation for the sense of self, the second of the two problems of consciousness I outlined in the introductory chapter—that is, how the owner of the movie-in-the-brain emerges within the movie. The proposal, however, does not fully address the first of the two problems I outlined then—how the movie-in-the-brain is generated from its qualic sources on up. Other proposals, from neurobiologists, cognitive scientists, and philosophers, are aimed at that first problem. For example, Gerald Edelman's proposal, perhaps the most comprehensive attempt to deal with the matter of consciousness published to date, uses an appealing biological framework to address the conditions under which the movie-in-the-brain can be generated.

In recent work, he carries the effort farther and specifies physiologic conditions necessary for the creation of integrated scenes in the conscious mind. Other thoughtful attempts to deal with aspects of the movie-in-the-brain problem include Bernard Baars's global workspace hypothesis and Daniel Dennett's multiple draft model.

Importantly, by making feelings be the primitives of consciousness, we are obliged to inquire about the intimate nature of feeling. What are feelings made of? What are feelings the perception of? How far behind feelings can we get? These questions are not entirely answerable at the moment. They define the edge of our current scientific reach.

Whatever the answers may turn out to be, however, the idea that human consciousness depends on feelings helps us confront the problem of creating conscious artifacts. Can we, with the assistance of advanced technology and neurobiological facts, create an artifact with consciousness? Perhaps not surprisingly, given the nature of the question, I have two answers for it, and one is no and the other yes. No, we have little chance of creating an artifact with anything that resembles human consciousness, conceptualized from an inner-sense perspective. Yes, we can create artifacts with the formal mechanisms of consciousness proposed in this book, and it may be possible to say that those artifacts have some kind of consciousness.

Some external behaviors of artifacts with formal mechanisms of consciousness will mimic conscious behaviors and may pass a consciousness version of the Turing test. But for all the good reasons that John Searle and Colin McGinn have adduced on the matter of behavior, mind, and the Turing test, passing the test guarantees little about the artifact's mind. More to the point, the artifact's internal states may even mimic some of the neural and mental designs I propose here as a basis for consciousness. They would have a way of generating second-order knowledge, but, without the help of the nonverbal vocabulary of feeling, the knowledge would not be expressed in the manner we encounter in humans and is probably present in so many living species. Feeling is, in effect, the barrier, because the realization of human consciousness may require the existence of feelings. The "looks" of emotion can be simulated, but what feelings feel like cannot be duplicated in silicon. Feelings cannot be duplicated unless flesh is duplicated, unless the brain's actions on flesh are duplicated, unless the brain's sensing of flesh after it has been acted upon by the brain is duplicated.

UNDER THE LIGHT

I began this book by invoking birth and the moment of stepping into the light as suggestive metaphors for consciousness. When self first comes to mind and forevermore after that, two-thirds of each living day without a pause, we step into the light of mind and we become known to ourselves. And now that the memory of so many becomings has created the persons we are, we can even imagine ourselves walking across the stage under the light.

It all begins modestly, with the barest of senses of our living being relating to some simple thing inside or outside the boundary of our bodies. Then the intensity of the light increases and as it gets brighter, more of the universe is illuminated. More objects of our past than ever before can be clearly seen, first separately, then at once; more objects of our future, and more objects in our surrounding are brightly lit. Under the growing light of consciousness, more gets to be known each day, more finely, and at the same time.

From its humble beginnings to its current estate, consciousness is a revelation of existence—a partial revelation, I must add. At some point in its development, with the help of memory, reasoning, and later, language, consciousness also becomes a means to modify existence.

All human creation comes back to that point of transition when we begin manipulating existence guided by the partial revelation of that very existence. We only create a sense of good and evil as well as norms of conscionable behavior once we know about our own nature and that of others like us. Creativity itself—the ability to generate new ideas and artifacts—requires more than consciousness can ever provide. It requires abundant fact and skill memory, abundant working memory, fine reasoning ability, language. But consciousness is ever present in the process of creativity, not only because its light is indispensable, but because the nature of its revelations guide the process of creation, in one way or another, more or less intensely. In a curious way, whatever we do invent, from norms of ethics and law to music and literature to science and technology, is either directly mandated or inspired by the revelations of existence that consciousness offers us. Moreover, in one way or another, more so or less, the inventions have an effect on existence as revealed, they alter it for better or for worse. There is a circle of influence—existence, consciousness, creativity—and the circle closes.

The drama of the human condition comes solely from consciousness. Of course, consciousness and its revelations allow us to create a better life for self and others, but the price we pay for that better life is high. It is not just the price of risk and danger and pain. It is the price of *knowing* risk, danger, and pain. Worse even: it is the price of knowing what pleasure is and *knowing* when it is missing or unattainable.

The drama of the human condition thus comes from consciousness because it concerns knowledge obtained in a bargain that none of us struck: the cost of a better existence is the loss of innocence about that very existence. The feeling of what happens is the answer to a question we never asked, and it is also the coin in a Faustian bargain that we could never have negotiated. Nature did it for us.

But drama is not necessarily tragedy. To some extent, in a variety of imperfect ways, individually and collectively, we have the means to guide creativity and, in so doing, improve human existence rather than worsen it. This is not easy to achieve; there are no blueprints to follow; the successes may be small; failure is likely. And yet, if creativity is directed successfully, even modestly, we will allow consciousness, once again, to fulfill its homeostatic, regulating role over existence. Knowing will help being. I even have some hope that understanding the biology of human nature will help a little with the choices to be made. Be that as it may, improving the lot of existence is precisely what civilization, the main consequence of consciousness, has been all about, and for at least three thousand years, with greater or smaller rewards, improvement is what civilization has been attempting. The good news, then, is that we have already begun.

Appendix

Notes on Mind and Brain

A GLOSSARY OF SORTS

Because words such as *images, neural patterns, representations,* and *maps* have unclear and various meanings, their use is fraught with difficulties. Nonetheless such words are indispensable to convey one's ideas in any attempt to deal with the topics of this book. These notes are meant to clarify further my usage of some of those words.

What Is an Image and What Is a Neural Pattern?
When I use the term *image*, I always mean *mental* image. A synonym for image is *mental pattern.* I do not use the word image to refer to the pattern of neural activities that can be found, with current neuroscience methods, in activated sensory cortices — for instance, in the auditory cortices in correspondence with an auditory percept; or in the visual cortices in correspondence with a visual percept. When I refer to the neural aspect of the process I use terms such as *neural pattern* or *map.*

Images can be conscious or nonconscious (see pages ahead). Nonconscious images are never accessible directly. Conscious images can be accessed *only in a first-person perspective* (my images, your images). Neural patterns, on the other hand, can be accessed *only in a third-person perspective*. If I had the chance of looking at my own neural patterns with the help of the most advanced technologies, I would still be looking at them from a third-person perspective.

Images Are Not Just Visual

By the term images I mean mental patterns with a structure built with the tokens of each of the sensory modalities—visual, auditory, olfactory, gustatory, and somatosensory. The somatosensory modality (the word comes from the Greek *soma* which means "body") includes varied forms of sense: touch, muscular, temperature, pain, visceral, and vestibular. The word image does not refer to "visual" image alone, and there is nothing static about images either. The word also refers to sound images such as those caused by music or the wind, and to the somatosensory images that Einstein used in his mental problem solving—in his insightful account, he called those patterns "muscular" images.[1] Images in all modalities "depict" processes and entities of all kinds, concrete as well as abstract. Images also "depict" the physical properties of entities and, sometimes sketchily, sometimes not, the spatial and temporal relationships among entities, as well as their actions. In short, the process we come to know as mind when mental images become ours as a result of consciousness is a continuous flow of images many of which turn out to be logically interrelated. The flow moves forward in time, speedily or slowly, orderly or jumpily, and on occasion it moves along not just one sequence but several. Sometimes the sequences are concurrent, sometimes convergent and divergent, sometimes they are superposed. *Thought* is an acceptable word to denote such a flow of images.

Constructing Images

Images are constructed either when we engage objects, from persons and places to toothaches, from the outside of the brain toward its inside; or when we reconstruct objects from memory, from the inside out, as it were. The business of making images never stops while we are awake and it even continues during part of our sleep, when we dream. One might argue that images are the currency of our minds. The words I am using to bring these ideas to you are first formed, however briefly and sketchily, as auditory, visual, or somatosensory images of phonemes and morphemes, before I implement them on the page in their written version. Likewise, those written words now printed before your eyes are first processed by you as verbal images before they promote the activation of yet other images, this time non-verbal, with which the "concepts" that correspond to my words can be displayed mentally. In this perspective, any symbol you can think of is an image, and there may be little leftover mental residue that is not made of images. Even the feelings that make up the backdrop of each mental instant are images, in the sense articulated above, so-matosensory images, that is, which mostly signal aspects of the body state. The obsessively repeated feelings that constitute the self in the act of knowing are no exception.

Images may be conscious or unconscious. It should be noted, however, that not all the images the brain constructs are made conscious. There are simply too many images being generated and too much competition for the relatively small window of mind in which images can be made conscious—the window, that is, in which images are accompanied by a sense that we are apprehending them and that, as a consequence, are properly attended. In other words, metaphorically speaking, there is indeed a subterranean underneath the conscious mind and there are many levels to that subterranean. One level is made of images not attended, the phenomenon to which I have just alluded. Another level is made of the neural patterns and of the relationships among neural patterns which subtend all images, whether they eventually become conscious or not. Yet another level has to do with the neural machinery required to hold records of neural patterns in memory, the kind of neural machinery which embodies innate and acquired implicit dispositions.

Representations

The meaning of a few other terms needs to be clarified. One is *representation,* a problematic but virtually inevitable term in discussions of this sort. I use representation either as a synonym of mental image or as a synonym of neural pattern. My mental image of a particular face is a representation, and so are the neural patterns that arise during the perceptual-motor processing of that face, in a variety of visual, somatosensory, and motor regions of the brain. This use of *representation* is conventional and transparent. It simply means "pattern that is consistently related to something," whether with respect to a mental image or to a coherent set of neural activities within a specific brain region. The problem with the term representation is not its ambiguity, since everyone can guess what it means, but the implication that, somehow, the mental image or the neural pattern *represents,* in mind and in brain, with some degree of fidelity, the object to which the representation refers, as if the structure of the object were replicated in the representation. When I use the word representation, I make no such suggestion. I do not have any idea about how faithful neural patterns and mental images are, relative to the objects to which they refer. Moreover, whatever the fidelity may be, neural patterns and the corresponding mental images are as much creations of the brain as they are products of the external reality that prompts their creation. When you and I look at an object outside ourselves, we form comparable images in our respective brains. We know this well because you and I can describe the object in very similar ways, down to fine details. But that does not mean that the image we see is the copy of whatever the object outside is like. Whatever it is like, in absolute terms, we do not know. The image we see is based on changes which occurred in our organisms—including the part of the organism called brain—when the physical structure of the object interacts with the body. The signaling devices located throughout our body structure—in the skin, in the muscles, in the retina, and so on—help construct neural patterns which map the organism's *interaction* with the object. The neural patterns are constructed according to the brain's own conventions, and are achieved transiently in the multiple sensory and motor regions of the brain that are suitable to process signals coming from particular body sites, say, the skin, or the muscles, or the retina. The building of those neural patterns or maps is based on the momentary selection of neurons and circuits engaged by the interaction. In other words, the building blocks exist within the brain, available to be picked up and assembled. The part of the pattern that remains in memory is built according to the same principles.

Thus the images you and I see in our minds are not facsimiles of the particular object, but rather images of the interactions between each of us and an object which engaged our organisms, constructed in neural pattern form according to the organism's design. The object is real, the interactions are real, and the images are as real as anything can be. And yet, the structure and properties in the image we end up seeing are brain constructions prompted by an object. There is no picture of the object being transferred from the object to the retina and from the retina to the brain. There is, rather, a set of correspondences between physical characteristics of the object and modes of reaction of the organism according to which an internally generated image is constructed. And since you and I are similar enough biologically to construct a similar enough image of the same thing, we can accept without protest the conventional idea that we have formed *the* picture of some particular thing. But we did not.

One final reason to be cautious about the term representation is that it easily conjures up the metaphor of the brain as computer. The metaphor is inadequate, however. The brain does perform computations but its organization and working have little resemblance to the common notion of what a computer is.

Maps

Many of the same qualifications apply to the term *map,* a word that is almost as inevitable and irresistible as *representation* when it comes to discussions on the neurobiology of the mind. When the light particles known as photons strike the retina in a particular pattern related to an object, the nerve cells activated in that pattern—say, a circle or a cross—constitute a transient neural "map." At subsequent levels of the nervous system, for instance, the visual cortices, subsequent related maps are also formed.[2] To be sure, just as with the word representation, there is a legitimate notion of pattern, and of correspondence between what is mapped and the map. But the correspondence is not point-to-point, and thus the map need not be faithful. The brain is a creative system. Rather than mirroring the environment around it, as an engineered information-processing device would, each brain constructs maps of that environment using its own parameters and internal design, and thus creates a world unique to the class of brains comparably designed.

Mysteries and Gaps of Knowledge in the Making of Images

There is no mystery regarding the question of where images come from. Images come from the activity of brains and those brains are part of living organisms that interact with physical, biological, and social environments. Accordingly, images arise from neural patterns, or neural maps, formed in populations of nerve cells, or neurons, that constitute circuits, or networks. There is a mystery, however, regarding *how* images emerge from neural patterns. How a neural pattern *becomes* an image is a problem that neurobiology has not yet resolved.

Many of us in neuroscience are guided by one goal and one hope: to provide, eventually, a comprehensive explanation for how the sort of neural pattern that we can currently describe with the tools of neurobiology, from molecules to systems, ever becomes the multidimensional, space-and-time-integrated image we are experiencing this very moment. The day may come when we can explain satisfactorily all the steps that intervene from neural pattern to image but that day is not here yet. When I say that images *depend on* and *arise from* neural patterns or neural maps, rather than saying they *are* neural patterns or maps, I am not slipping into inadvertent dualism, i.e., neural pattern, on one side, and nonmaterial cogitum, on the other. I am simply saying that we cannot characterize yet all the biological phenomena that take place between (*a*) our current description of a neural pattern, at varied neural levels, and (*b*) our experience of the image that originated in the activity within the neural map. There is a gap between our knowledge of neural events, at molecular, cellular, and system levels, on the one hand, and the mental image whose mechanisms of appearance we wish to understand. There is a gap to be filled by not yet identified but presumably identifiable physical phenomena. The size of the gap and the degree to which it is more or less likely to be bridged in the future is a matter for debate, of course. Be that as it may, I wish to make clear that I regard neural patterns as forerunners of the biological entities I call images.

The gap I have just described is one reason why, throughout this book, I maintain two levels of description, one for the mind and one for the brain. This separation is a simple matter of intellectual hygiene and, once again, it is not the result of dualism. By keeping separate levels of description I am not suggesting that there are separate substances, one mental and the other biological. I am simply recognizing the mind as a high level of biological process, which requires and deserves its own description because of the private nature of its appearance and because that appearance is the fundamental reality we wish to explain. On the other hand, describing neural events with their proper vocabulary is part of the effort to understand how those events contribute to the creation of the mind.

New Terms

Several new terms are introduced in this book, e.g., *core consciousness, extended consciousness* (which are first defined in chapter 1), and *proto-self* and *second-order structure* (which are properly introduced in chapters 5 and 6).

Also, my use of the terms *emotion* and *feeling* is unconventional, as I explain in the beginning of chapter 2; and the term *object* is used in a broad and abstract sense—a person, a place, and a tool are objects, but so are a specific pain or an emotion.

Some Pointers on the Anatomy
of the Nervous System

The nervous system is made of nervous or neural tissue. Like any other living tissue it is made of cells. The neural cells are known as *neurons* and although they are supported by another type of cell—the glial cell—everything indicates that neurons are the critical unit, the one unit essential to produce movements and mental activity.

Neurons have three main components: a *cell body*, the cell's power-house complete with cell nucleus and organelles such as mitochondria; a main output fiber known as the *axon;* and input fibers known as *dendrites.* Neurons are interconnected to form circuits in which one can find the equivalent of conducting wires (the neurons' axon fibers) and connectors, known as *synapses* (which usually consist of an axon making contact with the dendrites of another neuron).

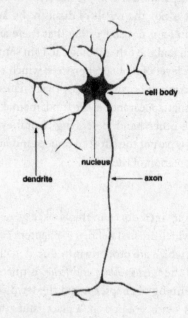

Figure A.1. A neuron and its main anatomical components.

There are billions of neurons in the human brain, organized in local circuits. Those circuits constitute *cortical regions,* if they are arranged in parallel layers, like a cake, or *nuclei,* if they are grouped in nonlayered collections, like berries in a bowl. Both the cortical regions and the nuclei are interconnected by axon "projections" to form *systems,* and, at gradually higher levels of complexity, *systems of systems.* When the axon projections are large enough to be individualized to the naked eye they form "pathways." In terms of scale, all neurons and local circuits are microscopic, while cortical regions, most nuclei, and systems are macroscopic.

For the purposes of anatomical description, the nervous system is usually divided into central and peripheral divisions. The main component of the *central nervous system* is the *cerebrum,* which is made up of the left and right *cerebral hemispheres* joined by the *corpus callosum* (a thick collection of nerve fibers connecting left and right hemispheres bidirectionally). The central nervous system also encompasses deep nuclei such as: (*a*) the *basal ganglia;* (*b*) the *basal forebrain;* and (*c*) the *diencephalon* (a combination of the *thalamus* and the *hypothalamus*). The cerebrum is joined to the *spinal cord* by the *brain stem,* behind which you can find the *cerebellum* (see figure A.2)

The central nervous system is connected to every point of the body by nerves, which are bundles of axons originating in the cell body of neurons. The collection of all nerves connecting the central nervous system (brain, for short) with the periphery and vice versa constitutes the *peripheral nervous system.* Nerves transmit impulses from brain to body and from body to brain. The brain and the body are also interconnected chemically, by substances such as hormones which course in the bloodstream.

A section of the central nervous system, in any direction you may wish to slice it, easily reveals a difference between dark and pale sectors. The dark sectors are known as the *gray matter* (although their real color is more brown than gray), and the pale sectors are known as the *white matter* (which is not that white, either). The gray matter gets its

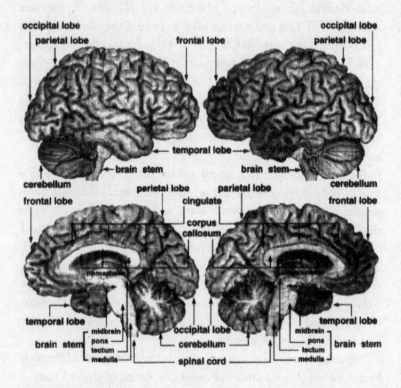

Right Hemisphere   **Left Hemisphere**

occipital lobe
parietal lobe
frontal lobe
occipital lobe
parietal lobe
temporal lobe
brain stem
brain stem
cerebellum
frontal lobe
parietal lobe
parietal lobe
cingulate
corpus callosum
cerebellum
frontal lobe
diencephalon
diencephalon
temporal lobe
midbrain
pons
tectum
medulla
brain stem
occipital lobe
cerebellum
spinal cord
temporal lobe
midbrain
pons
tectum
medulla
brain stem

Figure A.2. The main divisions of the central nervous system and their critical components, shown in 3-D reconstructions of a living human brain. The reconstructions are based on magnetic resonance data and on the BRAINVOX technique. Note the relative positions of the four principal lobes, of the diencephalon (which encompasses the thalamus and hypothalamus), and of the brain stem. Note also the position of the corpus callosum (which joins both hemispheres across the midline) and of the cingulate cortex of each hemisphere. The pattern of gyri and sulci is very similar in the left and the right cerebral hemispheres, but it is not equal: there are significant asymmetries and those asymmetries appear to underlie differences in function.

darker hue from the tight packing of massive numbers of neuron cell bodies. The nerve fibers, which emanate from the cell bodies located in the gray matter, constitute the white matter. The myelin sheath, which insulates the nerve fibers, gives the white matter its characteristic lighter appearance.

The gray matter comes in two varieties. Examples of the layered variety are the *cerebral cortex* which envelops the cerebral hemispheres, and the *cerebellar cortex* which envelops the cerebellum. Examples of the nonlayered variety, the *nuclei,* include: the *basal ganglia* (located in the depth of each cerebral hemisphere and made up of three large nuclei, the caudate, putamen, and pallidum); the *amygdala,* a single and sizable lump of nuclei located in the depth of each temporal lobe; and several aggregations of smaller nuclei which form the *thalamus,* the *hypothalamus,* and the gray sectors of the *brain stem.*

The cerebral cortex can be envisioned as a comprehensive mantle for the cerebrum, covering the surfaces of the cerebral hemisphere, including those that are located in the depths of fissures and sulci, the crevices which give the brain its characteristic folded appearance. The thickness of this multilayer mantle is about three millimeters, and the layers are parallel to each other and to the brain's surface. The evolutionarily modern part of the cerebral cortex is known as the *neocortex.* The cerebral cortex is an overwhelming presence, and all other gray structures, the various nuclei mentioned above, and the cerebellar cortex are known as subcortical. The main divisions of the cerebral cortex are designated as lobes: frontal, temporal, parietal, and occipital.

The various regions of the cortical lobes are traditionally identified by numbers corresponding to the distinctive architecture of its cellular arrangements (which is known as cytoarchitectonics). The numbering of the regions originated with the work of Korbinian Brodmann, and remains a valid tool after nearly a century. The numbers need to be learned, or checked in a map, and have nothing to do with the area's size or importance.

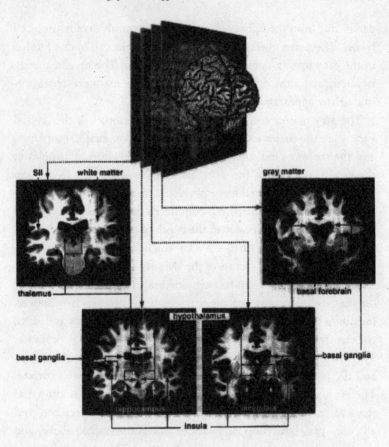

Figure A.3. Gray matter in the cerebral cortex and in deep nuclei. As noted in the text the gray matter is made up of densely packed cell bodies of neurons. The contrasting white matter contains the axons that originate in the cell bodies and travel to other regions in order to establish connections and transmit signals. The cross sections offer a view of the relative location of several deep structures not visible on the brain's surface—basal ganglia, basal forebrain, amygdala, thalamus, and hypothalamus. Note also the location of the insula, a region of the cortex that is part of the somatosensory system and is entirely hidden in the depth of the sylvian fissure.

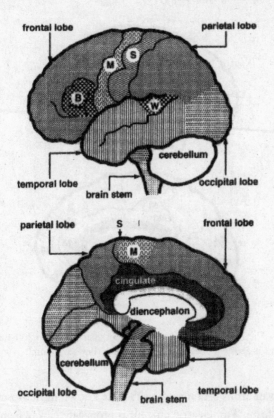

Figure A.4. The major anatomical regions of the cerebral hemispheres: frontal, temporal, parietal, and occipital lobes; Broca's (B) and Wernicke's (W) areas; motor (M) and somatosensory (S) areas. Although Broca's and Wernicke's areas are the best-known language-related brain regions, several other areas are also involved in language processing. Likewise for motor (M) and somatosensory (S) regions, which are just the tip of the motor and somatosensory icebergs. Elsewhere in the cerebral cortex, and underneath it, there are many cortical regions and nuclei that support motor function (cingulate cortex, basal ganglia, thalamus, brain-stem nuclei). The same applies to somatosensory function (brain-stem nuclei, thalamus, insula, cingulate cortex).

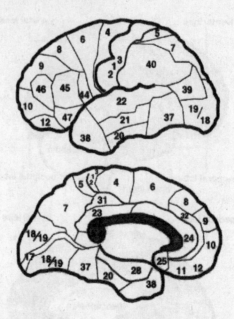

Figure A.5. The main Brodmann areas. The numbers do not reflect the function, the importance, or the location of these areas. They are a simple reference code.

When neurons become active (a state known in neuroscience jargon as "firing") an electric current is propagated away from the cell body and down the axon. When this current arrives at a synapse, it triggers the release of chemicals known as neurotransmitters (glutamate is an example of such a transmitter). In an excitatory neuron, the cooperative interaction of many other neurons whose synapses are adjacent determines whether or not the next neuron will fire, that is, whether it will produce its own action potential, which will lead to its own neurotransmitter release, and so forth.

Synapses can be strong or weak. Synaptic strength determines whether or not, and how easily, impulses continue to travel into the next neuron. In an excitatory neuron, a strong synapse facilitates impulse travel, while a weak synapse impedes or blocks it. On the average each neuron forms about 1,000 synapses. Considering that there are more than 10 billion neurons and more than 10 trillion synapses, each neuron tends to talk to a few others but never to most or all of the others. In fact, many neurons only talk to neurons that are not very far away, within relatively local circuits of cortical regions and nuclei, while others, although their axons travel for several centimeters, only make contact with a small number of other neurons. The action of neurons depends on the nearby assembly of neurons they belong to; whatever systems do depends on how assemblies influence other assemblies in an architecture of interconnected assemblies; and finally, whatever each assembly contributes to the function of the system to which it belongs, depends on its place in that system. The varied functions of different brain areas are a consequence of the place assumed by assemblies of sparsely connected neurons within large-scale systems. In short, the brain is a system of systems. Each system is composed of an elaborate interconnection of small but macroscopic cortical regions and subcortical nuclei, which are made of microscopic local circuits, which are made of neurons, all of which are connected by synapses.

The Brain Systems behind the Mind

For the purpose of investigating the relation between mental images and the brain, I have long used a framework suggested by results from experimental and clinical neuropsychology, neuroanatomy, and neurophysiology. The framework posits an *image space* and a *dispositional space*. The image space is that in which images of all sensory types occur explicitly. Some of those images constitute the manifest mental contents that consciousness lets us experience whereas some images remain nonconscious. The *dispositional space* is that in which dispositions contain the knowledge base and the mechanisms with which images can be constructed from recall, with which movements can be generated, and with which the processing of images can be facilitated. Unlike the contents of the image space, which are explicit, the contents of the dispositional space are implicit. We can know the contents of images (once core consciousness is activated), but we never know the contents of dispositions directly. The contents of dispositions are *always* nonconscious and exist in dormant form. Yet dispositions can produce a large variety of actions—the release of a hormone into the bloodstream; the contraction of muscles in viscera or of muscles in a limb or in the vocal apparatus. Dispositions hold some records for an image that was actually perceived on some previous occasion and participate in the attempt to reconstruct a similar image from memory. Dispositions also assist with the processing of a currently perceived image, for instance, by influencing the degree of attention accorded to the current image. We are never aware of the knowledge necessary to perform any of these tasks, nor are we ever aware of the intermediate steps that are taken. We are only aware of results, for example, a state of wellbeing; the racing of the heart; the movement of a hand; the fragment of a recalled sound; the edited version of the ongoing perception of a landscape.

All of our memory, inherited from evolution and available at birth, or acquired through learning thereafter, in short, all our memory of things, of properties of things, of persons and places, of events and relationships, of skills, of biological regulations, you name it, exists in dispositional form (a synonym for *implicit, covert, nonconscious*), waiting to become an explicit image or action. Note that dispositions are not words. They are abstract records of potentialities. Words or signs, which can signify any entity or event or relationship, along with the rules with which we put words and signs together also exist as dispositions and come to life as images and action, as in speech or signing. When I think of dispositions I always think of the town of Brigadoon waiting to come alive for a brief period.

We are beginning to discern which parts of the central nervous system support the image space and which parts support the dispositional space. The areas of cerebral cortex located in and around the arrival point of visual, auditory, and other sensory signals—the so-called *early sensory cortices* of the varied sensory modalities—support explicit neural patterns, and so do parts of limbic areas, such as the cingulate, and noncortical structures, such as the tectum. These neural patterns of maps continuously change under the influence of internal and external inputs and are likely to be the basis for images, whose mercurial dynamics parallel the neural pattern changes over time.

On the other hand, *higher-order cortices*—which make up the ocean of cerebral cortex around the islands of early sensory cortices and motor cortices—parts of limbic cortices, and numerous subcortical nuclei, from the amygdala to the brain stem, hold dispositions, that is, implicit records of knowledge. (See figure A.6.) When disposition circuits are activated they signal to other circuits and cause images or actions to be generated from elsewhere in the brain.

This bare sketch also requires the mention of other brain regions whose ostensible role is the interrelation of signals across brain areas, along with the control of their occurrence in certain brain areas. Those regions include the thalamus, the basal ganglia, the hippocampus, and the cerebellum. We would need a textbook to begin to discuss the intricacy of their respective jobs, in spite of the depth of our ignorance. For the sake of our discussion, however, I will simply say that the functions of the thalamus, e.g., interrelation of signals, control of brain activities in disparate areas, and relay of signals, are indispensable for consciousness. As far as consciousness goes, however, the role of the others is either unclear (basal ganglia, cerebellum) or negligible (hippocampus).

I have proposed that dispositions are held in neuron ensembles called convergence zones. To the partition between an image space and a dispositional space, then, corresponds a partition in (1) explicit neural pattern maps—activated in early sensory cortices, in so-called limbic cortices, and in some subcortical nuclei; and in (2) convergence zones, located in higher-order cortices and in some subcortical nuclei.

How this anatomical arrangement serves as a base for the sort of integrated and unified images we experience in our minds is not clear,

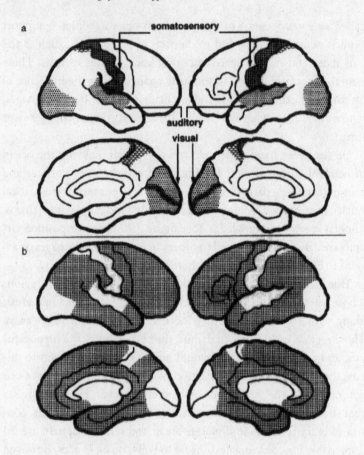

Figure A.6. **a** The main early sensory cortices (somatosensory, auditory, visual). The term "early" refers not to age in evolution but to order of entry of signals in the cerebral cortex. For example, light activates neurons in the retinas, then in the geniculate nuclei, and then in areas 17, 18, and 19, which are collectively known as "early visual cortices." Area 17 is also known as "primary visual cortex" or V_1. Areas 18 and 19 are also known as "visual association cortices," and they include subregions known as V_2, V_3, V_4, V_5. The same general arrangement applies to auditory and somatosensory cortices, respectively, in temporal and parietal lobes. **b** Higher-order and limbic cortices in crosshatched pattern. The remainder of the cerebral cortex is made up of higher-order cortices, which largely surround the early cortices, and of a few so-called limbic cortices, e.g., cingulate cortices.

although a number of proposals have suggested solutions to parts of this question. The question is generally known as the "binding" problem. In terms of an overall mental picture it is likely that binding requires some form of time-locking of neural activities that occur in separate but interconnected brain regions. There is little doubt that the integrated and unified scene that characterizes the conscious mind will require massive local and global signaling of populations of neurons across multiple brain regions. Gerald Edelman's notion of reentry addresses this requirement. Rodolfo Llinás's transcortical "binding wave" and my notion of time-locked retroactivation are other attempts to capture a mechanism capable of making the necessarily fragmented activity of our brain cohere in time and space.[3] The work of Wolf Singer has addressed the mechanisms required to generate coherence at the microstructural level,[4] and Francis Crick has theorized extensively about those requirements, at cellular and microcircuit levels.[5] Both Jean-Pierre Changeux and Gerald Edelman have proposed selectional frameworks for the operation of such mechanisms, and the work of Michael Merzenich shows that the brain does have the flexibility necessary to operate in this manner.[6]

ENDNOTES

Chapter One: Stepping into the Light

1 Consciousness has long been an important topic in philosophy but until recently only a few neuroscientists had worked on it. For a brief period during the middle of the twentieth century, especially in the forties and fifties, neuroscience devoted considerable attention to the study of consciousness. The experimental work of G. Magoun, H. W. Moruzzi, and H. Jasper and the clinical and experimental observations of W. Penfield stand out among several contributions from an epoch that ended all too soon. Benjamin Libet is another pioneering exception. What is currently known as the field of consciousness studies was created over the past decade by a handful of philosophers and scientists, independently, unwittingly, and unexpectedly. Thanks are due especially to the philosophers Daniel Dennett, Paul and Patricia Churchland, Thomas Nagel, Colin McGinn, and John Searle, and to the neuroscientists Gerald Edelman and Francis Crick.

2 I outlined the problem in chapter 10, "The Body-Minded Brain," of *Descartes' Error: Emotion, Reason, and the Human Brain* (New York: Putnam, 1994; Avon Hearst, 1995).

3 For a pertinent review see J. Levine, "Materialism and qualia: The explanatory gap," *Pacific Philosophical Quarterly* 64 (1983): 354–61.

4 See Daniel Dennett's *Consciousness Explained* (Boston: Little, Brown, 1991) for a comprehensive discussion of homunculus explanations for the sense of self.

5 Failure to distinguish the two problems of consciousness outlined in these pages leads to equivocal situations. For instance, I interpret the remarkable efforts of the mathematical physicist Roger Penrose as pertaining to the elucida-

tion of the physical basis of the qualia problem, although they are invariably described as pertaining to consciousness as a whole. The same applies to the work of the physicist Henry Stapp. Neither body of work focuses on the part of the problem of consciousness that I am emphasizing in this book but rather on the more general and by no means less-important problem of the biological basis of the mental process. See R. Penrose, *The Shadows of the Mind* (New York: Oxford University Press, 1994); and H. Stapp, *Mind, Matter, and Quantum Mechanics* (Berlin: Springer Verlag, 1993).

6 Given the scale of the challenge it should not be surprising that in the attempt to deal with the matter of consciousness, philosophers and neurobiologists, alike, face numerous barriers and are unlikely to find a comprehensive solution anytime soon. For example, the word *consciousness*, polygamously wed to far too many meanings, has often stood in the way of agreement regarding the definition of the problem; the private nature of the phenomenon has dissuaded many from even approaching the matter and has convinced others that it can be approached in a purely external manner, and privacy be damned; the notion that, somehow, consciousness stands at the very peak of human abilities has often given rise to a paralyzing awe and to the belief that consciousness is beyond our scientific reach; impatience and the desire to cut through the above impedimenta have led some to conclude that not only is consciousness approachable, it is already perfectly elucidated; finally, there are those who think the problem does not exist at all, or that it is nothing other than the problem of mind: consciousness can be elucidated or not depending on whether the problem of mind is or is not elucidated. Against this background my position is that the problem of consciousness exists and has not been solved yet; that it can be broken down in parts; that consensus can be generated regarding those parts; and that in spite of its private nature, consciousness can be approached scientifically.

7 The term *mind*, as I use it in this book, encompasses both conscious and nonconscious operations. It refers to a *process*, not a thing. What we know as mind, with the help of consciousness, is a continuous flow of mental patterns, many of which turn out to be logically interrelated. The flow moves forward in time, speedily or slowly, orderly or jumpily, and on occasion it moves along not just one sequence but several. Sometimes the sequences are concurrent, sometimes convergent and divergent, sometimes they are superposed.

The term I frequently use as shorthand for mental patterns is *images*. As noted, images are mental patterns in any sensory modality, not just visual. There are sound images, or tactile images, and so on.

8 There is no unanimity of views on the relation between mind and brain, especially

as it relates to consciousness. It is not possible to cite every author who has published recent and important texts on this general issue, but I recommend a number of books or collections by philosophers of mind who have given careful thought to these problems. Their positions and mine do not always coincide, but I have enjoyed reading all of the following: John Searle, *The Rediscovery of the Mind* (Cambridge, Mass.: MIT Press, 1992); Patricia and Paul Churchland, *On the Contrary* (Cambridge, Mass.: MIT Press, 1998); David J. Chalmers, *The Conscious Mind* (New York: Oxford University Press, 1996); Daniel Dennett, *Consciousness Explained* (cited earlier); Thomas Nagel, *The View from Nowhere* (New York: Oxford University Press, 1986); Colin McGinn, *The Problem of Consciousness* (Oxford: Basil Blackwell, 1991); Owen Flanagan, *Consciousness Reconsidered* (Cambridge, Mass.: MIT Press, 1992); Ned Block, Owen Flanagan, Güven Güzeldere, eds., *The Nature of Consciousness: Philosophical Debates* (Cambridge, Mass.: MIT Press, 1997); Thomas Metzinger, ed., *Conscious Experience* (Paderborn, Germany: Imprint Academic/Schöningh, 1995); Fernando Gil, *Modos de Evidência* (Lisbon: Imprensa Nacional, 1998); Jerry A. Fodor, *The Modularity of Mind* (Cambridge, Mass.: MIT Press, 1983).

9 H. Damasio, and A. Damasio, *Lesion Analysis in Neuropsychology* (New York: Oxford University Press, 1989).

10 The separation of consciousness into at least two levels of phenomena is well supported by cognitive and behavioral analyses and validated by the neurological observations I present here. The separation is a must when it comes to proposing biological mechanisms capable of producing consciousness. No single mechanism is likely to deliver both core and extended consciousness. This problem has also been identified in another biological account of consciousness, Gerald Edelman's. The dichotomy he proposes also separates the "simple" from the "complex," although his categories do not correspond to mine. Gerald Edelman divides consciousness into primary and higher-order consciousness, but his primary consciousness is simpler than my core consciousness and does not result in the emergence of a self. Edelman's higher-order consciousness is also not the same as my extended consciousness, because it requires language and is strictly human.

Other authors have proposed dichotomous classifications of consciousness. For instance, Ned Block divides consciousness into access, or A-consciousness, and phenomenal, or P-consciousness. Neither concept is related to the notions of core and extended consciousness. See Gerald Edelman, *The Remembered Present* (New York: Basic Books, 1989); Ned Block, et al., *The Nature of Consciousness* (cited earlier).

11 A consensus has been developing recently that subjectivity is the "hard problem" of consciousness, although discussions of subjectivity usually do not consider that it requires a subject—a sense of self—and that the means whereby

we have a sense of self, illusory or not, must be an important aspect of eluci-dating consciousness. The term "hard problem" was introduced to a large pub-lic by David Chalmers in *The Conscious Mind* (cited earlier) and is the latest designation for the old qualia problem. For an earlier statement of the problem see J. Levine, "Materialism and qualia" (cited earlier). For a recent discussion of this problem see John Searle, *The Mystery of Consciousness* (New York: New York Review of Books, 1997).

12 For an account of how the visual system achieves such object representations see David Hubel's *Eye, Brain and Vision* (New York: Scientific American Library, 1988) and Semir Zeki's *A Vision of the Brain* (Oxford: Blackwell Scientific Publica-tions, 1993).

13 B. Spinoza.*The Ethics*, Part IV, Proposition 22 (Indianapolis: Hackett Publishing Co., Inc., 1982, first published 1677).

Chapter Two: Emotion and Feeling

1 Ludwig von Bertalanffy, *Modern Theories of Development: An Introduction to Theoretical Biology* (New York: Harper, 1962, originally published in German in 1933); P. Weiss, "Cellular dynamics," *Review of Modern Physics* 31 (1919): 11–20; Kurt Goldstein, *The Organism* (New York: Zone Books, 1995, originally published in German in 1934).

2 See Gerald Edelman, *The Remembered Present* (cited earlier) and Antonio Damasio, *Descartes' Error*. Other notable exceptions: Theodore Bullock authored a biologi-cal text written from an evolutionary perspective, *Introduction to Nervous Systems* (San Francisco: W. H. Freeman, 1977); Paul MacLean talked about a triune brain, each of the three tiers belonging to an evolutionary age in "The Triune Brain, Emotion, and Scientific Bias," in *The Neurosciences: The Second Study Program*, ed. F. O. Schmitt (New York: Rockefeller University Press, 1970); and Patricia Churchland launched neurophilosophy with an invocation of the value of considering evo-lution in *Neurophilosophy: Toward a Unified Science of the Mind-Brain* (Cambridge, Mass.: MIT Press, Bradford Books, 1986).

3 Some of the examples of this change can be found in the work of Jean-Didier Vincent and Alain Prochiantz in France, Joseph LeDoux, Michael Davis, James McGaugh, Jerome Kagan, Richard Davidson, Jaak Panksepp, Ralph Adolphs, and Antoine Bechara in North America; and Raymond Dolan, Jeffrey Gray, and E. T. Rolls in Britain, to name but some of the most visible.

4 A. Damasio, *Descartes' Error* (cited earlier); A. Damasio, "The somatic marker hy-pothesis and the possible functions of the prefrontal cortex," *Philosophical Transac-tions of the Royal Society of London*, Series B (Biological Sciences) 351 (1996): 1413–20; A. Bechara, A. Damasio, H. Damasio, and S. Anderson, "Insensitivity to future consequences following damage to human prefrontal cortex," *Cognition* 50 (1994):

7–15; A. Bechara, D. Tranel, H. Damasio, and A. Damasio, "Failure to respond autonomically to anticipated future outcomes following damage to prefrontal cortex," *Certebral Cortex* 6 (1996): 215–25; A. Bechara, H. Damasio, D. Tranel, and A. Damasio, "Deciding advantageously before knowing the advantageous strategy," *Science* 275 (1997): 1293–95.

5 For a discussion on the cognition of rationality see N. S. Sutherland, *Irrationality: The Enemy Within* (London: Constable, 1992); for its cognitive and biological aspects see Patricia S. Churchland, "Feeling Reasons," in Paul M. Churchland and Patricia S. Churchland, *On the Contrary* (cited earlier).

6 Other languages that have conveyed the heritage of Western philosophy and psychology have long had available the equivalent of the separate English words *emotion* and *feeling*. For example: Latin: *exmovere* and *sentire;* French: *émotion* and *sentiment;* German: *Emotionen* and *Gefühl;* Portuguese: *emoção* and *sentimento;* Italian: *emozione* and *sentimento,* and so on. The two words were probably coined in those several languages because many clear-eyed observers, as they considered the two distinguishable sets of phenomena, sensed their separation and saw the value of denoting them by different terms. Referring to the whole process by the single word *emotion,* as is now common practice, is pure carelessness. Nor should it be forgotten that in its more general meaning the word *feeling* denotes perceptions related to the body—feelings of malaise or well-being, feelings of pain, the feeling of something touched—rather than an appreciation of what is seen or heard. The wise coiners of the word *feeling* were probably under the correct impression that feeling an emotion had a lot to do with the body, and they were right on the mark.

7 D. Tranel and A. Damasio, "The covert learning of affective valence does not require structures in hippocampal system or amygdala," *Journal of Cognitive Neuroscience* 5 (1993): 79–88.

8 There is also evidence from studies in healthy individuals, without brain lesions, that preferences can be learned nonconsciously and quite rapidly. See P. Lewicki, T. Hill, and M. Czyzewska, "Nonconscious acquisition of information," *American Psychologist* 47 (1992): 796–801, for a specific experiment. For reviews of this area of studies, see J. Kihlstrom, "The cognitive unconscious," *Science* 237 (1987): 285–94; Arthur S. Reber, *Implicit Learning and Tacit Knowledge: An Essay on the Cognitive Unconscious* (New York: Oxford University Press, 1993).

9 Deciding what constitutes an emotion is not an easy task, and once you survey the whole range of possible phenomena, one does wonder if any sensible definition of emotion can be formulated, and if a single term remains useful to describe all these states. Others have struggled with the same problem and con-

cluded that it is hopeless. See Leslie Brothers, *Friday's Footprint: How Society Shapes the Human Mind* (New York: Oxford University Press, 1997) and Paul Griffiths, *What Emotions Really Are: The Problem of Psychological Categories* (Chicago: University of Chicago Press, 1997). At this point, however, my preference is to retain the traditional nomenclature, clarify the use of the terms, and wait until further evidence dictates a new classification, my hope being that by maintaining some continuity we will facilitate communication at this transitional stage. I will talk about three levels of emotion—background, primary, and secondary. This is revolutionary enough for one day, given that background emotions are not part of the usual roster of emotions. I will refer to drives and motivations and pain and pleasure as triggers or constituents of emotions, but not as emotions in the proper sense. No doubt all of these devices are intended to regulate life, but it is arguable that emotions are more complex than drives and motivations, than pain and pleasure.

10 Emotions have varied temporal profiles. Some emotions tend to be engaged in a "burst" pattern. They go through a fairly rapid onset, a peak of intensity, and rapid decay. Anger, fear, surprise, and disgust are cases in point. Other emotions have more of a "wavelike" pattern; some forms of sadness and all of the background emotions are prime examples. It should be clear that many variations of profile are possible depending on circumstances and on individuals.

When states of emotion tend to become fairly frequent or even continuous over long periods of time, it is preferable to refer to them as *moods* rather than emotions. I believe moods should be distinguished from background emotions; a particular background emotion can be sustained over time to create a mood. If people think you are "moody" it is because you have been sounding a prevailing emotional note (perhaps related to sadness or anxiety) consistently for a good part of the time or maybe you have changed your emotional tune unexpectedly and frequently. Fifty years ago you would have been called "neurotic," but nobody is neurotic anymore.

Moods can be pathological, and we speak, then, of mood disorders. Depression and mania are the standard examples. You are depressed when the emotion sadness is dragged on for days, weeks, and months, when melancholic thoughts and crying and loss of appetite, sleep, and energy are not a single burst or a gentle wave but a continuous mode of being, physically and mentally. The same applies to mania. It is one thing to jump with joy at the right event or to be enthusiastic about your prospects in life, it is another to sustain the joy and exuberance on for days on end, justifiably or not. For powerful descriptions of the experience of mood disorders see Kay Redfield Jamieson, *An Unquiet Mind* (New York: Knopf, 1995); William Styron, *Darkness Visible: A Memoir of Madness* (New York:

Random House, 1990); and Stuart Sutherland, *Breakdown: A Personal Crisis and a Medical Dilemma,* updated edition (London: Weidenfeld and Nicolson, 1987). See Robert Robinson for medical information on mood disorders.

Because moods are dragged-out emotions along with the consequent feelings, moods carry over time the collections of responses that characterize emotions: endocrine changes, autonomic nervous system changes, musculoskeletal changes, and changes in the mode of processing of images. When this whole package of reactions is deployed persistently and inappropriately over long periods of time, the cost to the individual so affected is prohibitive. The term *affect* is often used as a synonym of "mood" or "emotion" although it is more general and can designate the whole subject matter we are discussing here: emotions, moods, feelings. Affect is the thing you display (emote) or experience (feel) toward an object or situation, any day of your life whether you are moody or not.

11 The critical differences between "background" emotions and the "conventional" emotions thus rest with: (1) the source of the immediate inducer, which is usually external or representing the exterior in the case of "conventional" emotions, and is internal in the case of background emotions; and (2) the focus of the responses, whose targets favor the musculoskeletal and visceral systems in "conventional" emotions but favor the internal milieu in "background" emotions. The entire evolution of emotions must have begun with background emotions. When we compare background emotions to "the big six" and to the so-called "social" emotions, we note a progressive degree of specificity of the inducers, of specificity of the responses, and of specificity of the response targets, a progressive differentiation of controls, from global to local.

12 P. Ekman, "Facial expressions of emotions: New findings, new questions," *Psychological Science* 3 (1992): 34–38.

13 The terms "social" or "secondary" should not suggest that these emotions are solely generated by education within a culture. In an interesting essay on the emotions, Paul Griffiths (*What Emotions Really Are,* cited earlier) notes, correctly, that secondary emotions are not the result of culture alone; this made me realize that I did not emphasize this idea strongly enough in *Descartes' Error.* No doubt the role played by society in the shaping of secondary emotions is greater than in the case of the primary emotions. Moreover, it is clear that several "secondary" emotions begin to appear later in human development, probably only after a concept of self begins to mature—shame and guilt are examples of this later development; newborns have no shame and no guilt but two-year-olds do. That does not mean, however, that secondary emotions are not biologically preset, in part or mostly.

14 R. Bandler and M. T. Shipley, "Columnar organization in the midbrain peri-aqueductal gray: Modules for emotional expression?" *Trends in Neurosciences* 17 (1994): 379–89; M. M. Behbehani, "Functional characteristics of the midbrain periaqueductal gray," *Progress in Neurobiology* 46 (1995): 575–605; J. F. Bernard and R. Bandler, "Parallel circuits for emotional coping behaviour: New pieces in the puzzle," *Journal of Comparative Neurology* 401 (1998): 429–46.

15 A. Damasio, T. Grabowski, H. Damasio, A. Bechara, L. L. Ponto, and R. Hichwa, "Neural correlates of the experience of emotion," *Society for Neuroscience* Abstracts 24 (1998): 258. Our finding of brain-stem activation in negative emotions is novel, and so is the finding of hypothalamic activation in sadness. Our finding of activation in ventromedial prefrontal cortex confirms previous findings of M. E. Raichle, J. V. Pardo, and P. J. Pardo; E. M. Reiman, R. Lane, and colleagues; and Helen Mayberg.

16 See Joseph LeDoux, *The Emotional Brain: The Mysterious Underpinnings of Emotional Life* (New York: Simon and Schuster, 1996) for a review of animal research in the topic of fear.

17 M. Mishkin, "Memory in monkeys severely impaired by combined but not separate removal of amygdala and hippocampus," *Nature* 273 (1978): 297–98; Larry Squire, *Memory and Brain* (New York: Oxford University Press, 1987); F. K. D. Nahm, H. Damasio, D. Tranel, and A. Damasio, "Cross-modal associations and the human amygdala," *Neuropsychologia* 31 (1993): 727–44; Leslie Brothers, *Friday's Footprint* (cited earlier).

18 A. Bechara, D. Tranel, H. Damasio, R. Adolphs, C. Rockland, and A. R. Damasio, "A double dissociation of conditioning and declarative knowledge relative to the amygdala and hippocampus in humans," *Science* 269 (1995): 1115–18.

19 R. Adolphs, D. Tranel, and A. R. Damasio, "Impaired recognition of emotion in facial expressions following bilateral damage to the human amygdala," *Nature* 372 (1994): 669–72. R. Adolphs, H. Damasio, D. Tranel, and A. R. Damasio, "Cortical systems for the recognition of emotion in facial expressions," *Journal of Neuroscience* 16 (1996): 7678–87.

20 R. Adolphs, and A. R. Damasio, "The human amygdala in social judgement," *Nature* 393 (1998): 470–74.

21 Curiously, when the brain mechanisms underlying emotion are compromised, the ability to attribute emotion to the simple chip is impaired. This is what Andrea Heberlein and Ralph Adolphs have just shown in our laboratory. Patients with damage to specific emotion induction sites describe the shapes and movements of the chips in an accurate, matter-of-fact manner. Spontaneously, however, they fail to assign emotions to the chips or to their interrelations. The manifest intellectual level of the show is perceived without flaw

but the emotional subtext is not detected. A. S. Heberlein, R. Adolphs, D. Tranel, D. Kemmerer, S. Anderson, and A. Damasio, "Impaired attribution of social meanings to abstract dynamic visual patterns following damage to the amygdala," *Society for Neuroscience* Abstracts 24 (1998): 1176.

22 Eric R. Kandel, Jerome Schwartz, and Thomas M. Jessell, eds., *Principles of Neural Science,* 3rd ed. (Norwalk, Conn.: Appleton and Lange, 1991).

23 I have previously described this episode in *Descartes' Error,* and I will briefly summarize it here.

24 P. Rainville, G. H. Duncan, D. D. Price, B. Carrier, and M. C. Bushnell, "Pain affect encoded in human anterior cingulate but not somatosensory cortex," *Science* 277 (1997): 968–71; P. Rainville, R. K. Hofbauer, T. Paus, G. H. Duncan, M. C. Bushnell, and D. D. Price, "Cerebral mechanisms of hypnotic induction and suggestion," *Journal of Cognitive Neuroscience* 11 (1999): 110–25; P. Rainville, B. Carrier, R. K. Hofbauer, M. C. Bushnell, and G. H. Duncan, "Dissociation of pain sensory and affective dimensions using hypnotic modulation," *Pain* (in press).

25 See A. K. Johnson and R. L. Thunhorst, "The neuroendocrinology of thirst and salt appetite: Visceral sensory signals and mechanisms of central integration," *Frontiers in Neuroendocrinology* 18 (1997): 292–353, for a review of the complex mechanisms involved in behaviors such as thirst.

Chapter Three: Core Consciousness

1 John Searle has presented a lucid defense of this position in *The Rediscovery of the Mind* (cited earlier). Daniel Dennett has argued similarly in *Consciousness Explained* (cited earlier).

2 The descriptions of coma and vegetative state are presented in chapter 8 and are well covered in textbooks of neurology. A standard reference is the text by Fred Plum and Jerome B. Posner, a classic volume in which they survey their unique experience in the neurology of coma. See F. Plum and J. B. Posner, *The Diagnosis of Stupor and Coma,* 3rd ed. (Philadelphia: F. A. Davis Company, 1980).

3 Jean-Dominique Bauby, *Le scaphandre et le papillon* (Paris: Editions Robert Laffont, 1997); J. Mozersky, *Locked In: A Young Woman's Battle with Stroke* (Toronto: The Golden Dog Press, 1996).

4 The descriptions of epileptic states and of akinetic mutism are standard, and can be found in numerous articles and textbooks of neurology. Accessible references include: Wilder Penfield and Herbert Jasper, *Epilepsy and the Functional Anatomy of the Human Brain* (Boston: Little, Brown, 1954); J. Kiffin Penry, R. Porter, and F. Dreifuss, "Simultaneous recording of absence seizures with video tape and electroencephalography, a study of 374 seizures in 48 patients," *Brain* 98 (1975): 427–40; F. Plum and J. B. Posner, *The Diagnosis of Stupor and Coma* (cited earlier); and

A. Damasio and G. W. Van Hoesen, "Emotional disturbances associated with focal lesions of the limbic frontal lobe," in *The Neuropsychology of Human Emotion: Recent Advances,* ed. Kenneth Heilman and Paul Satz (New York: The Guilford Press, 1983): 85-110. The inferences I draw on the standard evidence, as far as consciousness is concerned, are based on my own observations of patients so affected.

5 I discuss this evidence in chapter 5, in the context of outlining the representation of objects.

6 My comments on epilepsy and emotion pertain to the situation of absence seizures. When automatisms occur in the setting of so-called temporal lobe seizures, emotions can appear before or during the episode. Partial impairments of emotion are not associated with abolition of core consciousness. For instance, the patients with ventromedial frontal lobe lesions described in *Descartes' Error* only lose secondary emotions. They lose their ability to react with embarrassment in a social situation or to react with fear in relation to a possible financial loss in the distant future, but most of their background emotions and primary emotions remain in place. Likewise, as we saw in the discussion of patient S, damage to the amygdala impairs some primary and secondary emoting related to fear, but not other primary and secondary emotions, and it does not compromise background emotions at all.

Chapter Four: The Hint Half Hinted

1 This is an issue that deserves further attention. I have encountered few exceptions to the observation that impaired core consciousness goes with impaired emotion, but it would be important to study the exceptions systematically. In my experience, they consist largely of "shamlike" bursts of rage or laughter, i.e., unmotivated behaviors suggestive of release of automated routines, and they occur in persistent vegetative state or in nonabsence seizures associated with temporal lobe damage.

2 The work of Francis Crick exemplifies this position. Inasmuch as a comprehensive elucidation of consciousness requires an understanding of the process of image making, Crick's approach is fruitful—we certainly must understand how the brain comes to form images, and his hypotheses offer several testing opportunities. But Crick believes that "there are many forms of consciousness, such as those associated with seeing, thinking, emotion, pain and so on" and that "self-consciousness—that is, the self-referential aspect of consciousness—is probably a special case of consciousness. In our view, it is better left to one side for the moment." F. Crick, *The Astonishing Hypothesis: The Scientific Search for the Soul* (New York: Scribner, 1994). My concern is that the elimination of self-reference may create a barrier to the comprehensive solution of the problem of consciousness.

3 In a critical and important review Güven Güzeldere lists a handful of "inner sense" contemporary philosophers: David Armstrong, Paul Churchland, Daniel Dennett, David Rosenthal, Peter Carruthers, and William Lycan. See G. Güzeldere, "Is Consciousness the Perception of What Passes in One's Mind?", in T. Metzinger ed., *Conscious Experience* (cited previously, see chapter 1).

4 Consciousness is selective because it does not encompass all objects in mind. To put it simply, some objects can become more conscious than others. In the welter of images of objects that might be made conscious, not all are. The truth is that not every object is equal because some objects are more valuable than others for an organism concerned with maintaining life.

Consciousness is a continuous property of the mind because in normal and awake minds, things to be known are continuously being represented. This is a consequence of the condition of complex awake organisms: they are either perceptually engaged with the outside world or busily producing internally recalled images, or most commonly, both. Whether the machinery that generates consciousness does so discretely rather than continuously is a different matter. I believe the machinery actually produces "pulses" of core consciousness, many singular units of consciousness occurring one after the other from several consciousness generators. The interval between units is so small and the amount of parallel pulses so abundant, that we only register a continuous, whirring blur.

Consciousness pertains to objects other than itself. There is an object, on the one hand, and there is consciousness of the object, on the other, separable from it although clearly related to it. Consciousness is "other" than the objects about which it is, a critical separation that is often ignored in modern accounts of consciousness.

Consciousness is personal in that it arises in a given organism and is about events in that organism. By "personal" James also meant that it was internal, unobservable to outsiders. The properties of consciousness I outlined earlier provide a description for the components of that last and paramount property: the personal aspect of consciousness. *Individual perspective* helps define the personal nature of Jamesian consciousness. *Individual ownership* completes the definition of the personal and so does *individual agency.* See William James, *The Principles of Psychology,* vol. 1 (New York: Dover Publications, 1950).

5 B. Libet, "Timing of cerebral processes relative to concomitant conscious experience in man," in *Advances in Physiological Sciences,* ed. G. Adam, I. Meszaros, and E. I. Banyai (Elmsford, N.Y.: Pergamon Press, 1981).

6 The neuropsychologist Marc Jeannerod has shown that the process of executing motor activity effectively masks the mental process that constitutes the preparation of the movements. See M. Jeannerod, "The representing brains: Neural

correlates of motor intention and imagery," *Behavioural Brain Sciences* 17 (1994): 187-202. The neurophysiologist Alain Berthoz has studied the underlying physiology in detail. See A. Berthoz, *Les sens du mouvement* (Paris: Editions Odile Jacob, 1997).

Chapter Five: The Organism and the Object

1 *Descartes' Error,* chapter 10 and introduction.

2 Claude Bernard, *Introduction a l'étude de la médecine expérimentale* (Paris: J. B. Baillière et fils, 1865); Walter B. Cannon, *The Wisdom of the Body* (New York: W. W. Norton and Co., 1932).

3 Steven Rose, *Lifelines: Biology beyond Determinism* (New York: Oxford University Press, 1998).

4 In searching for precedents to the general idea that somehow the body is a basis for the self, I have encountered it in Kant, Nietzsche, Freud, and Merleau-Ponty, although not in the manner in which I articulate the idea with the tripartite arrangement of proto-self, core self, and autobiographical self, and not with the emphasis on homeodynamic stability. Edelman's distinction between self and nonself is also based on a body/nonbody distinction, although in his framework the self refers to biological individuality and does not connect with the conscious self of my proposal in the same manner. The philosophers Mark Johnson and George Lakoff establish a close connection between cognition and body representation, and so does the neurophysiologist Nicholas Humphrey. Israel Rosenfield also links body and self, but indirectly, through memory, his sense of self being weighted toward the sort of self I call autobiographical.

5 F. Nietzsche, in the prologue to *Thus Spake Zarathustra.* Some translations refer to "phantom" for "ghost," and "disharmony" for "discord."

6 This aspect of biology is often surprisingly ignored. For an exception I recommend Humberto Maturana and Francisco Varela, two biologists who have coined an appropriate word to describe the rebuilding process of living cells: *autopoiesis.* See H. Maturana and F. Varela, *The Tree of Knowledge: The Biological Roots of Human Understanding,* rev. ed. (Boston: Shambhala, 1992). In general, these notions have a counterpart in the philosophy of Alfred North Whitehead. See A. N. Whitehead, *Process and Reality* (New York: Free Press, 1969 c. 1929). On a related note, Pierre Rainville called my attention to the notion of "neuromatrix" developed by Ronald Melzack in connection with his studies of pain and phantom limbs. Melzack suggests that we are born with a genetically controlled neural network, modifiable by experience, which supports our feeling of the body. This would explain why many children born without limbs do experience "phantoms" for the arms and hands they never had. It would also help

explain some of the phantom limb phenomena recently studied by V. S. Ramachandran.

7   For a review on the mechanisms that permit us to make perceptuomotor adjustments, see the work of Alain Berthoz (previously cited).

8   The fact that "senses" are naturally combined recalls the notion of synesthesia. Synesthesia is a rare phenomenon. In the few individuals who have it, it tends to wane or disappear after childhood. It consists of perceiving a stimulus in a given sensory modality, for instance, a sound, and having the stimulus provoke a related experience, for instance, a color or smell. The differentiation of our non-synesthetic sensory devices usually prevents us from apprehending sensory signals in blended form; those who have the creative twist of real synesthesia apprehend the intermingling of the senses directly. Synesthetes tend to develop consistent linkages between certain sensations, e.g., a musical note and a number. Several brilliant composers and musical prodigies have been synesthetes and some nineteenth-century thinkers had the intriguing intuition that synesthesia might hold a key to the understanding of consciousness. They were not far from a hot trail, I may add. The Russian neuropsychologist A. R. Luria offered a rich description of synesthesia in his description of the mnemonist Solomon S., a case dramatized by Peter Brook and Marie-Hélène Estiènne in their play *"Je suis un phenomène!"* and movingly staged by Brook at the Théatre des Bouffes-du-Nord. Richard Cytowic has written a valuable review on synesthesia; see *The Man Who Tasted Shapes* (New York: Putnam, 1993).

9   A. Craig, "An ascending general homeostatic afferent pathway originating in lamina I," in *Progress in Brain Research* 107 (1996): 225–42; Z. Han, E. T. Zhang, and A. D. Craig, "Nociceptive and thermoreceptive lamina I neurons are anatomically distinct," *Nature Neuroscience* 1 (1998): 218–25.

10 W. D. Willis and R. E. Coggeshall, *Sensory Mechanisms of the Spinal Cord,* 2nd ed. (New York: Plenum Press, 1991). See also Craig (1996), cited above, for a thoughtful discussion of the integration of "body" senses at different levels of the nervous system, from the spinal cord to the cerebral cortex.

11 Coming at the problem from a very different perspective, the philosopher Fernando Gil has advanced the concept of a similarly nonconscious precursor entity and baptized it with the same name. We had never talked about the problem and we discovered the compatibility of our views on the same day and same place, hearing our respective lectures.

    The term self is used widely in disciplines such as immunology and psychology, and the meanings vary considerably although the notion of single individual is shared by all those usages. The psychological literature contains insightful discussions on the notion of self, for instance, Ulric Neisser's discussion of five selves (although none of them correspond to the levels I describe, and although, unlike mine, all of them are based on external rather than internal information). In the neurobiological literature, Gerald Edelman's "concept of self" corresponds to the upper reaches of my autobiographical self. See U. Neisser, "Five kinds of self-knowledge," *Philosophical Psychology* 1 (1988): 35–59; G. Edelman, *The Remembered Past* (cited earlier, see chapter 1).

12 See the discussion on the concept of reticular formation in chapter 8.

13 J. Panksepp, *Journal of Consciousness Studies* 5 (1998): 566–82. In a related development, in the fall of 1998, Douglas Watt posted an essay on the Internet in which he related emotion to consciousness. Efforts such as Panksepp's and Watt's are both rare and welcome.

14 G. Tononi, O. Sporns, and G. Edelman provide a plausible model for the sort of interactions required by such a process within early sensory cortices; see "Reentry and the problem of integrating multiple cortical areas: Simulation of dynamic integration in the visual system," *Cerebral Cortex* 2 (1992): 310–35. In a recent article, G. Tononi and G. Edelman expand that model substantially so that it can encompass large-scale cortical integration; see "Neuroscience: Consciousness and complexity," *Science* 282 (1998): 1846–51.

15 A. Damasio, "Time-locked multiregional retroactivation," 1989; A. Damasio, "The brain binds entities and events," 1989 (cited earlier).

16 A. Damasio, D. Tranel, and H. Damasio, "Face agnosia and the neural substrates of memory," *Annual Review of Neuroscience,* 13 (1990): 89–109.

17 D. Tranel, A. Damasio, and H. Damasio, "Intact recognition of facial expression, gender, and age in patients with impaired recognition of face identity," *Neurology* 38 (1988): 690–96.

18 A. Damasio, H. Damasio, and G. Van Hoesen, "Prosopagnosia: Anatomic basis and behavioral mechanisms," *Neurology* 32 (1982): 331–41.

19 N. Kanwisher, J. McDermott, and M. M. Chun, "The fusiform face area: A module in human extrastriate cortex specialized for face perception," *Journal of Neuroscience* 17 (1997): 4302–11.

20 D. K. Meno, A. M. Owen, E. J. Williams, P. S. Minhas, C. M. C. Allen, S. J. Boniface, J. D. Pickard, I. V. Kendall, S. P. M. J. Downer, J. C. Clark, T. A. Carpenter, and N. Antoun, "Cortical processing in persistent vegetative state," *Lancet* 352 (1998): 800. This interesting finding should not be taken to mean that all patients with persistent vegetative state will show such patterns of activation. Because of the extent and distribution of their lesions some patients will not.

Chapter Six: The Making of Core Consciousness

1 An example may help clarify the idea further. Consider a situation in which a concrete object is actually present in front of an organism and is being apprehended

through vision. I will deal later with the situation in which objects are present in recall although the essence of the process is not different.

The critical events that occur in our organism when we confront an object are of two principal sorts. First, there are changes in the organism's state caused by adjustments required by the perceptuomotor process, e.g., eye movements, head and body movements, hand movements, vestibular changes, and so forth. Second, there are changes caused by the impact of the object on the state of internal milieu and viscera. The latter includes the sort of responses which eventually generate emotions and which begin to change both the organism and its representation even before actual emotional states occur. We should remember here that our previous experience with both a specific object and with the same kind of object turns virtually any object into an inducer of some emotional reaction, weak or strong, good, bad, or in between. We should also remember that, as I noted earlier, emotion has a truly dual status in relation to consciousness: The actual responses whose consequences, as an ensemble, eventually produce an emotion are part of the mechanism that drives core consciousness; a fraction of time later, however, the collections of responses which constitute a particular emotion can also be treated as an object to be known. When the "emotional" object is made conscious, it becomes a feeling of emotion.

From the brain's standpoint the critical events described above are signaled in the specific regions appropriate to signal the object and the proto-self, as previously discussed. However, the nonverbal account that I propose as the critical component of consciousness is based on yet *other* brain structures and describes how the events I just enumerated are *caused* by the ongoing sensory representation of the object's presence and by the organism's *obligate* reaction to it, in mechanical terms and in terms of emotional value. The nonverbal account establishes the relationship between object, on the one hand, and organism, as represented by the proto-self, on the other. It tells a clear story—a primordial story—and the secret of the plot is that the organism has been changed by the object.

2 The words "description," "caused," and "relationship" in these sentences mean precisely what they appear to mean. By *description* I mean mapped neural signals; *caused* and *relationship* pertain to the close temporal succession between the occurrence of the object images and the occurrence of the accompanying images. I do not mean to say that the brain is pre-equipped to detect causality. Causality and logical relationships possibly arise naturally in the processes realized by a brain with a particular anatomy. In the same vein, the brain does not need a prior sense of "objectness," although the design of the brain's perceptual systems and the different significance of varied objects for the organism's welfare do help carve out objects from the welter of stimuli impinging on the organism's somatomotor apparatus.

3 You may wonder if the nonverbal account I just described is a fiction, and if knowing and self are illusions. That is an interesting question and it has more than one
answer, but my answer is that they are not fictional. After all, we do come to verify independently, a posteriori, in our beings and in other beings, that the kinds of
characters in the primordial plot, e.g., the living individual organisms, the objects, and the relationships portrayed in the plot, are in fact consistent, systematic, and widespread occurrences. In that sense they are not fictional because they
respect a relative truth standard. On the other hand, it is difficult to imagine that
they depict any absolute truth. At the scale of the universe, the achievement of
consciousness is modest and what it lets us see is limited.

4 A. Damasio and H. Damasio, "Cortical systems for retrieval of concrete knowledge: The convergence zone framework," in *Large-Scale Neuronal Theories of the
Brain,* ed. Cristof Koch and Joel L. Davis (Cambridge, Mass.: MIT Press, 1994):
61–74; A. Damasio, "Concepts in the brain," *Mind and Language* 4 (1989): 24–28.

5 Jerome Kagan, *The Second Year: The Emergence of Self-Awareness* (Cambridge, Mass.:
Harvard University Press, 1981); M. Lewis, "Self-conscious emotions," *American
Scientist* 83 (1995): 68–78.

6 For more background on the process of recall see *Descartes' Error* (chapter 9) and
Daniel Schacter's *Searching for Memory: The Brain, the Mind, and the Past* (New York:
Basic Books, 1996). My notion of recall is based on Frederic Bartlett who introduced the idea that we do not recall facsimiles of perceived objects, but rather
reconstruct, as best we can, some approximation to the original perception.
Frederic C. Bartlett, *Remembering: A Study in Experimental and Social Psychology* (Cambridge, England: The University Press, 1954).

7 John Ashbery, "Self-Portrait in a Convex Mirror," in *Selected Poems* (New York:
Penguin, 1986).

8 R. W. Sperry, M. S. Gazzaniga, and J. E. Bogen, "Interhemispheric relationships:
The neocortical commissures; syndromes of their disconnection," in *Handbook of
Clinical Neurology,* ed. P. J. Vinken and G. W. Bruyn, vol. 4 (Amsterdam: North-
Holland, 1969): 273–90.

9 Julian Jaynes, *The Origin of Consciousness in the Breakdown of the Bicameral Mind* (Boston:
Houghton Mifflin, 1976); D. Dennett, *Consciousness Explained* (cited earlier); H. Maturana and F. Varela, *The Tree of Knowledge* (cited earlier).

10 The line appears motivated by a nonexceptional event—a guard alone in the
night inquires, "Who's there?" when he hears footsteps. Yet, this is no mere "*qui
vive*" and there is little chance that Shakespeare did not use it deliberately as a
means to announce the deep inquiry of his play. Some years ago Peter Brook
exposed the importance of this inaugural question in a play he wrote and
staged, based on *Hamlet* and titled *Qui est là?*

11 Others have commented, directly or indirectly, on the existence of a storytelling

stance in the human mind. Daniel Dennett, when he describes his multiple-drafts model of consciousness, is using verbal storytelling implicitly as a base for what I call *extended consciousness*. Michael Gazzaniga has called attention to the fabulist language tendencies of the human left cerebral hemisphere in split-brain patients, and postulated a language-based cortical "interpreter"; and Mark Turner has suggested that literary narratives are homologues for higher-cognitive processes. See D. Dennett, *Consciousness Explained* (cited earlier); M. Gazzaniga, *The Mind's Past* (Berkeley: University of California Press, 1998); and M. Turner, *The Literary Mind* (New York: Oxford University Press, 1996).

12 A notion reminiscent of my second-order map has been discussed by Wolf Singer (1998) and by Gerd Sommerhoff (1996). In both instances, they see the need to form metarepresentations of ongoing brain activities, but the neural site for the representations is quite different from mine in the case of Singer (who proposes to locate it in newer cortical structures such as the prefrontal cortices) and is not specified in the case of Sommerhoff. In both instances, the result of the metarepresentations would be some kind of global workspace rather than a sense of self as I specify. W. Singer, "Consciousness and the structure of neuronal representations," *Philosophical Transactions of the Royal Society of London* Series B (Biological Sciences) 353 (1998): 1829–40; G. Sommerhoff, "Consciousness Explained as an Internal Integrating System," *Journal of Conscious Studies* 3 (1996): 139–57.

Chapter Seven: Extended Consciousness

1 Jerome Kagan, *The Second Year* (cited earlier); M. Lewis, "Self-conscious emotions," 1995 (cited earlier).

2 See P. Goldman-Rakic, "Circuitry of primate prefrontal cortex and regulation of behavior by representational memory," in *Handbook of Physiology: The Nervous System*, vol. 5, ed. F. Plum and V. Mountcastle (Bethesda, Md.: American Physiological Society, 1987): 353–417; A. Baddeley, "Working memory," *Science* 255 (1992): 566–69; Edward Smith and John Jonides for references on working memory in general (E. E. Smith, J. Jonides, and R. A. Koeppe, "Dissociating verbal and spatial working memory using PET," *Cerebral Cortex* 6 [1996]: 11–20; E. E. Smith, J. Jonides, R. A. Koeppe, E. Awh, E. H. Schumacher, and S. Minoshima, "Spatial versus object working-memory: PET investigations," *Journal of Cognitive Neuroscience* 7 [1995]: 337–56); and Stanislas Dehaene and Jean-Pierre Changeux for a proposed connection between working memory and consciousness (at Gulbenkian Symposium on Consciousness, 1998).

3 Bernard J. Baars, *A Cognitive Theory of Consciousness* (New York: Cambridge University Press, 1988). See also J. Newman, "Putting the puzzle together, Part II: To-

wards a general theory of the neural correlates of consciousness," *Journal of Consciousness Studies* 4:2 (1997): 100–21.

4 Hans Kummer, *In Quest of the Sacred Baboon: A Scientist's Journey* (Princeton, N.J.: Princeton University Press, 1995); Marc D. Hauser, *The Evolution of Communication* (Cambridge, Mass.: MIT Press, 1996).

5 A. Damasio, N. R. Graff-Radford, H. Damasio, "Transient partial amnesia," *Archives of Neurology* 40 (1983): 656–57.

6 J. Babinski, "Contribution à l'étude des troubles mentaux dans l'hémiplégie organique cérébrale (anosognosie)," *Revue neurologique* 27 (1914): 845–47.

7 The oblivion that anosognosic patients express toward their sick limbs is matched by the lack of concern they show for their overall situation. The news that there was a major stroke and that the likelihood of further and severe health problems is high is usually greeted with equanimity. On the contrary, when you give comparable bad news to a patient with the mirror-image damage in the left hemisphere, the reaction is entirely normal. In a systematic study of anosognosic patients, my colleague Steven Anderson has confirmed that anosognosia extends beyond paralysis to encompass the entire health situation of the patient and its implications. Having a defective autobiography deprived of proper updating, patients with anosognosia cannot construct an adequate theory for what is happening now, for what may happen in the future, and for what others think of them. They are also unaware that their theorizing is inadequate. When the autobiographical self-image is so compromised, it is no longer possible to realize that the thoughts and actions of that self are no longer normal. See S. Anderson and D. Tranel, "Awareness of disease states following cerebral infarction, dementia, and head trauma: Standardized assessment," *The Clinical Neuropsychologist* 3 (1989): 327–39.

8 One should ask why this map is skewed to the right hemisphere rather than being bilateral, considering that the body has two almost symmetrical halves. The answer: in humans as well as nonhuman species, functions seem to be apportioned asymmetrically to the cerebral hemispheres, probably because one final controller is better than two when it comes to choosing an action or a thought. (If both sides had equal say on making a movement, you might end up with a conflict; your right half might interfere with the left, and you would have a lesser chance of producing coordinated patterns of motion involving more than one limb.) For some functions, structures in one hemisphere must have an advantage, a functional arrangement known as dominance.

The best-known example of dominance concerns language. (In more than 95 percent of all people, including many left-handers, language depends largely on left-hemisphere structures.) Another example of dominance, this one favoring the right hemisphere, involves integrated body sense. As noted earlier, this is not a single, continuous map but rather a set of coordinated separate maps. The representation of extrapersonal space, the higher level of representation of body state, and the representation of emotion, all involve a right-hemisphere dominance.

9 Kenneth Heilman has recently added an interesting aspect to this traditional view by suggesting that the patients also lack an intention to move and are thus robbed of a means to check their own defect easily. K. M. Heilman, A. M. Barrett, and J. C. Adair, "Possible mechanisms of anosognosia: a defect in self-awareness," *Philosophical Transactions of the Royal Society of London* Series B (Biological Science series) 353 (1998): 1903–9.

10 A. Damasio, "Time-locked multiregional retroactivation," 1989; A. Damasio, "The brain binds entities," 1989 (cited earlier); A. Damasio and H. Damasio, "Cortical systems for retrieval of concrete knowledge," in *Large-Scale Neuronal Theories of the Brain* (cited earlier).

11 For a discussion of the neural basis for concepts and respective words, see: H. Damasio, T. J. Grabowski, D. Tranel, R. D. Hichwa, and A. Damasio, "A neural basis for lexical retrieval," *Nature* 380 (1996): 499–505; D. Tranel, H. Damasio, and A. Damasio; "A neural basis for the retrieval of conceptual knowledge," *Neuropsychologia* 35 (1997): 1319–27; D. Tranel, C. G. Logan, R. J. Frank, and A. Damasio, "Explaining category-related effects in the retrieval of conceptual and lexical knowledge for concrete entities: Operationalization and analysis of factors," *Neuropsychologia* 35 (1997): 1329–39.

12 Daniel Dennett, *Consciousness Explained* (cited earlier).

13 Alfred North Whitehead, *Process and Reality, Part 3* (New York: The Free Press, 1978, c. 1929).

14 The framework that I am presenting for the autobiographical self lends itself to thinking about so-called multiple personalities in neurobiological terms. In these strange and controversial cases, patients seem to switch from one particular identity and set of personal characteristics to another, and in some cases there are more than two identities. The switch is not as sharp as that depicted in *The Three Faces of Eve* (book and film), and it appears the culture of this condition and the therapeutic milieu in which the patients are immersed have a lot to do with the shape of the clinical presentation. Nonetheless, something unusual does occur in these patients that goes beyond the limits of acceptable character transformation in most of us. (See Ian Hacking, *Rewriting the Soul: Multiple Personalities and the Sciences of Memory* [Princeton, N.J.: Princeton University Press, 1995].) It is possible that instead of having one single set of rallying points for the generation of identity and personality, that is, one single set of intercon-

nected convergence zones/dispositions for one single identity and personality connected with one single organism, such individuals manage to create, because of varied circumstances of their past history, more than one master control site. I suspect the multiple master control sites are located in the temporal and frontal cortices and that the switch from one master control to another enables the identity/personality switch to occur. The switch would involve thalamic coordination, as in the instance of a normal single personality. In such patients, to a certain extent, it is reasonable to talk about more than one "autobiographical memory," and more than one construction of identity and response manner, connected to different life histories and anticipated futures. It is apparent, however, that in spite of being able to display more than one autobiographical self, such patients continue to have only one mechanism of core consciousness and only one core self. Each of the autobiographical selves must use the same central resource. Reflection on this fact is intriguing. It brings us back to the notion that the generation of the core self is closely related to the proto-self which, in turn, is based closely on the representations of one singular body in its singular brain. Given a single set of representations for one body state, it would require a major pathological distortion to generate more than one proto-self and more than one core self. Presumably the distortion would not be compatible with life. On the other hand, the generation of the autobiographical self occurs at a higher anatomical and functional level, no doubt connected to the core self, but partially independent of it and therefore less influenced by the strong biological shadow of a singular organism.

The distinction between the highly constrained organization of the core self, tied to biological organization in an inevitable manner, and the organization of autobiographical memory, potentially removed from biological constraints by some degrees of freedom, underscores the different degrees of allegiance to nature and to nurture, of, respectively, core self and autobiographical self. Curiously, in keeping with this idea, there is evidence that although multiple personalities may be linked to certain types of biological propensity, they are highly dependent on cultural factors for their development and shaping.

15 *"Gott, welch Dunkel hier!"* Ludwig von Beethoven, *Fidelio,* act 2, scene 1.

16 D. Schacter, 1996, ibid.; A. Damasio, D. Tranel, and H. Damasio, "Face agnosia and the neural substrates of memory," *Annual Review of Neuroscience* 13 (1990): 89–109.

17 E. R. Dobbs, *The Greeks and the Irrational* (Berkeley: University of California Press, 1951).

18 J. Jaynes, *The Origin of Consciousness* (cited earlier).

19 Kathleen Wilkes has written an interesting essay on the word consciousness which complements the differences I cover here by discussing how languages

such as Chinese and Hungarian cope with the concept. See K. V. Wilkes, "—, yishi, duh, um, and consciousness," in *Consciousness in Contemporary Science*, ed. A. J. Marcel and E. Bisiach (Oxford: Clarendon Press, 1992): 16–41.

20 Jean-Pierre Changeux, *Fondements naturels de l'éthique* (Paris: Editions Odile Jacob, 1993); J.-P. Changeux, *Une même ethique pour tous?* (Paris: Editions Odile Jacob, 1997); J.-P. Changeux and Paul Ricoeur, *Ce qui nous fait penser: La nature et la règle* (Paris: Editions Odile Jacob, 1998); D. Dennett, *Consciousness Explained* (cited earlier); B. Baars, *A Cognitive Theory of Consciousness* (cited earlier); J. Newman, "Putting the puzzle together," 1997; Robert Ornstein, *The Evolution of Consciousness* (New York: Prentice-Hall, 1991); Robert Ornstein and Paul Ehrlich, *New World, New Mind* (New York: Simon and Schuster, Touchstone, 1989).

Chapter Eight: The Neurology of Consciousness

1 F. Plum and J. Posner, *The Diagnosis of Stupor and Coma* (cited earlier) is a recommended reference for further reading on the subject.

2 The idea that neurologists have formed from cases of coma and vegetative state (that consciousness has been disrupted to the very core and that the mind is suspended for all intents and purposes) is just as clear to lay observers and shows up in popular culture. The film *Reversal of Fortune* provides a good example. In Nicholas Kazan's script, the film traces the events that led to Sunny von Bulow's coma and persistent vegetative state. Shortly after it begins there is a shot of the perfectly still body of Sunny (played by Glenn Close) accompanied by her voiceover telling us that she is no longer conscious or capable of behaving! "Brain dead, body better than ever," she says. The public immediately grasps its gallows humor absurdity. To have a comatose character narrate its state for the viewer is just one step removed from the even more absurd notion of having a dead character report on the events that led to death in the first place. Incidentally, that is precisely what Billy Wilder had his character Joe Gillis do in his remarkable *Sunset Boulevard*. At the start of the film, the very dead Gillis (played by William Holden) quietly floats, facedown, in Gloria Swanson's swimming pool and begins telling the audience, in voiceover, how he came to be shot and killed. That these dramatic devices succeed so well and are so memorable indicates the degree to which the central notions of what consciousness is and is not have been absorbed by the nonspecialists.

3 Ann B. Butler and William Hodos, "The reticular formation," in *Comparative Vertebrate Neuroanatomy: Evolution and Adaptation* (New York: Wiley-Liss, Inc., 1996): 164–79.

4 Coma and persistent vegetative state can also be caused by extensive bilateral damage to the thalamus or by widespread bilateral damage to the cerebral cortex.

Coma and persistent vegetative state are caused, more often than not, by structural brain damage as opposed to metabolic changes. Common causes of such damage are cerebral vascular disease, which leads to a stroke, and head injury, which produces results similar to a stroke in the sense that, by direct mechanical injury or by injury to blood vessels, brain tissue ends up collapsing. There can be other causes of these conditions, however, and there are interesting interrelations between coma and persistent vegetative state outlined below.

When coma occurs as a result of structural damage, caused by a stroke or by head injury, the location of the damage is as indicated in the previous section: there is damage to the upper half of the brain-stem tegmentum at high pontine level and/or at midbrain level, and the hypothalamus is often damaged as well. But coma can also be caused by damage to particular nuclei in the thalamus, namely, the intralaminar nuclei. The latter nuclei are part of the long upward pathway that originates in brain stem and is eventually disseminated throughout the cerebral cortex. Note that in all of these instances of structural damage, it is necessary to damage *both* the left and right sides of the structure. Unilateral damage of the critical areas does not alter consciousness.

5 For an example of the sort of interaction that can occur among such nuclei, see G. Aston-Jones, M. Ennis, V. A. Pieribone, W. T. Nickell, and M. T. Shipley, "The brain nucleus locus coerulus: Restricted afferent control of a broad efferent network," *Science* 234 (1986): 734–7; and B. E. Van Bockstaele and G. Aston-Jones, "Integration in the ventral medulla and coordination of sympathetic, pain, and arousal functions," *Clinical and Experimental Hypertension* 17 (1995): 153–65.

6 Carlo Loeb and John Stirling Meyer, *Strokes Due to Vertebro-Basilar Disease; Infarction, Vascular Insufficiency, and Hemorrhage of the Brain Stem and Cerebellum* (Springfield, Ill.: Charles C. Thomas, 1965): 188; R. Fincham, T. Yamada, D. Schottelius, S. Hayreh, and A. Damasio, "Electroencephalographic absence status with minimal behavior change," *Archives of Neurology* 36 (1979): 176–78.

7 Locked-in syndrome is commonly caused by structural damage in the anterior aspect of the pons and midbrain, as outlined above, but it can be caused by a severe polyneuropathy, a situation in which the nerves carrying signals necessary for the contraction of muscles are so dysfunctional that there is a widespread paralysis. Certain drugs can also mimic the locked-in condition. A drug known as curare, which blocks the nicotinic receptors of acetylcholine that are necessary for nerve fibers to command muscular contraction, results in widespread paralysis of the muscles under voluntary control. Contraction of smooth (nonstriated) muscles depends on a different kind of receptor, the muscarinic receptor, and thus curare does not block neuromuscular transmission to those receptors. As a result, the nonvoluntary commands to alter the caliber of blood

vessels or to modify the state of several viscera, which occur in emotion and in plain homeostatic regulation, can still take place in a fully curarized individual.

8   F. Plum and J. Posner, *The Diagnosis of Stupor and Coma* (cited earlier).

9   A. B. Scheibel and M. E. Scheibel, "Structural substrates for integrative patterns in the brainstem reticular core," in *Reticular Formation of the Brain*, ed. H. Jasper, L. D. Proctor, R. S. Knighton, D. C. Noshy, and R. T. Costello (Boston: Little, Brown, 1958): 31–55.

10 Alf Brodal, *The Reticular Formation of the Brain Stem: Anatomical Aspects and Functional Correlations* (Edinburgh: The William Ramsay Henderson Trust, 1959); J. Olszewski, "Cytoarchitecture of the human reticular formation," in *Brain Mechanisms and Consciousness*, ed. J. F. Delafresnaye, et al. (Springfield, Ill.: Charles C. Thomas, 1954): 54–80; W. Blessing, "Inadequate frameworks for understanding bodily homeostasis," *Trends in Neurosciences*, 20 (1997): 235–39.

11 J. Allan Hobson, *The Chemistry of Conscious States: How the Brain Changes Its Mind* (New York: Basic Books, 1994).

12 G. Moruzzi and H. W. Magoun, "Brain stem reticular formation and activation of the EEG," *Electroencephalography and Clinical Neurophysiology* 1 (1949): 455–73; F. Bremer, "Cerveau 'isolé' et physiologie du sommeil," *C. R. Soc. Biol.* 118 (1935): 1235–41.

13 R. Llinás and D. Paré, "Of dreaming and wakefulness," *Neuroscience* 44 (1991): 521–35; M. Steriade, "New vistas on the morphology, chemical transmitters, and physiological actions of the ascending brainstem reticular system," *Archives Italiennes de Biologie* 126 (1988): 225–38; M. Steriade, "Basic mechanisms of sleep generation," *Neurology* 42 (1992): 9–17; M. Steriade, "Central core modulation of spontaneous oscillations and sensory transmission in thalamocortical systems," *Current Opinion in Neurobiology* 3 (1993): 619–25; M. Steriade, "Brain activation, then (1949) and now: Coherent fast rhythms in corticothalamic networks," *Archives Italiennes de Biologie* 134 (1995): 5–20; M. H. J. Munk, P. R. Roelfsema, P. Koenig, A. K. Engel, and W. Singer, "Role of reticular activation in the modulation of intracortical synchronization," *Science* 272 (1996): 271–74; J. A. Hobson, *The Chemistry of Conscious States* (cited earlier); R. Llinás and U. Ribary, "Coherent 40-Hz oscillation characterizes dream state in humans," *Proceedings of the National Academy of Sciences of the United States*, 90 (1993): 2078–81.

14 For a review on the anatomy of acetylcholine nuclei see M. Mesulam, C. Geula, M. Bothwell, and L. Hersh, "Human reticular formation: Cholinergic neurons of the pedunculopontine and laterodorsal tegmental nuclei and some cytochemical comparisons to forebrain cholinergic neurons," *The Journal of Comparative Neurology* 283 (1989): 611–33. For general reviews on monoamine systems, see F. E. Bloom, "What is the role of general activating systems in cortical function?" in *Neurobiology of Neocortex*, ed. P. Rakic and W. Singer (New York: John Wiley & Sons Limited, 1997): 407–21; R. Y. Moore, "The reticular formation: monoamine

neuron systems," in *The Reticular Formation Revisited: Specifying Function for a Nonspecific System,* ed. J. A. Hobson and M. A. B. Brazier (New York: Raven Press, 1980): 67–81.

15 This is not the place to review these interesting findings although some references are provided should the reader wish to pursue this matter. See A. J. Hobson, *The Chemistry of Conscious States;* M. Steriade, "Basic mechanisms of sleep generation," 1992.

16 M. H. J. Munk, et al., "Role of reticular activation," 1996; M. Steriade, "Arousal: revisiting the reticular activating system," *Science* 272 (1996): 225–26.

17 M. Steriade and M. Deschenes, "The thalamus as a neuronal oscillator," *Brain Research* 320 (1984): 1–63. Also see J. E. Bogen, for a pertinent review, "On the neurophysiology of consciousness: 1. An overview," *Consciousness and Cognition,* 4 (1995): 52–62.

18 D. A. McCormick and M. von Krosigk, "Corticothalamic activation modulates thalamic firing through glutamate 'metabotropic' receptors," *Proceedings of the National Academy of Sciences of the United States* 89 (1992): 2774–8; R. Llinás and D. Paré, "Of dreaming and wakefulness" (cited earlier), 1991.

19 A. Brodal, *The Reticular Formation of the Brain Stem* (cited earlier).

20 F. Bremer, "Cerveau 'isolé' et physiologie du sommeil" (cited earlier).

21 C. Batini, G. Moruzzi, M. Palestini, G. Rossi, and A. Zanchetti, "Persistent pattern of wakefulness in the pretrigeminal midpontine preparation," *Science* 128 (1958): 30–32.

22 Another relevant experiment in relation to the first prediction concerns a study in cats, performed almost four decades ago by Sprague and colleagues (J. M. Sprague, M. Levitt, K. Robson, C. N. Liu, E. Stellar, and W. W. Chambers, "A neuroanatomical and behavioral analysis of the syndromes resulting from midbrain lemniscal and reticular lesions in the cat," *Archives Italiennes de Biologie* 101 [1963]: 225–95). The investigators damaged the ascending sensory tracts on one or the other side of the upper brain stem, and in some cases on both sides. The unilateral cases are interesting in themselves, but I will comment only on the bilateral cases. As a result of the lesions all the somatosensory inputs describing body state were cut and thus unavailable to the upper midbrain, hypothalamus, thalamus, and cerebral cortex. The lesions also interrupted auditory and vestibular inputs. The reticular nuclei of the lower and middle brain stem, however, continued to receive somatosensory signals although it is likely that at least some signals from cerebral cortex aimed at reticular nuclei were also blocked by the lesions. The result of these lesions was a profound change in behavior marked by an abolition of emotionality; by neglect of olfactory stimuli (which enter the brain at a higher level, directly into cerebral cortex); and by aimless, stereotypical behaviors, unrelated to the stimuli in the surroundings and to the animals' needs. Sprague and his colleagues described the animals in a most suggestive way, saying that they looked like automata. They were awake

but were deprived of emotion and were unconnected with the situation. They remained so for two and a half years, until they were sacrificed for the purpose of postmortem study.

The suggestions and questions raised by this study are fascinating. In the very least, the study suggests that intact reticular nuclei can generate wakefulness and permit behavior, but do not guarantee the sort of appropriate and adaptive behavior which hallmarks the presence of consciousness and planning. The study also suggests that a continuous diet of current-body-state signals must be necessary to maintain emotion, and, in all likelihood, consciousness. This suggestion needs to be tempered, in part, by the possibility that damage to pathways from cortex to reticular nuclei occurred and contributed to the defect, although it is not reasonable to presume that damage to downward cortical input alone could explain the results. Finally, I must point out the similarity between some of the behaviors noted in the cats and the presentation of patients with the partial disorders of consciousness that I described earlier, e.g., epileptic automatisms. Wakefulness is present but the behaviors are stereotypical, not part of a sensible context-related plan, and there is no evidence that a core consciousness and a core self are being formed.

For those interested in the history of neuroscience, I should add that this experiment led Sprague to investigate the role of the superior colliculus in vision. Sprague noted that the lesions he had produced had unintentionally undercut the superior colliculi. All of those cats had the abnormalities noted above as well as visual neglect. In the one cat in which the lesions did not undercut the colliculus, all the abnormalities were still present but visual neglect was missing (J. M. Sprague, in *The History of Neuroscience in Autobiography*, ed. L. R. Squire [Washington, D.C.: Society for Neuroscience, 1996]).

23 M. H. J. Munk, P. R. Roelfsema, P. Koenig, A. K. Engel, and W. Singer, "Role of reticular activation in the modulation of intracortical synchronization," *Science* 272 (1996): 271–74.

24 S. Kinomura, J. Larsson, B. Gulyás, and P. E. Roland, "Activation by attention of the human reticular formation and thalamic intralaminar nuclei," *Science* 271 (1996): 512–15.

25 R. Bandler and M. T. Shipley, "Columnar organization in the midbrain periaqueductal gray," *Trends in Neurosciences* 17 (1994): 379–89; M. M. Behbehani, "Functional characteristics of the midbrain periaqueductal gray," *Progress in Neurology* 46 (1995): 575–605; J. F. Bernard and R. Bandler, "Parallel circuits for emotional coping behavior," *J. Comp. Neurol.* 401 (1998): 429–36.

26 J. Parvizi, G. W. Van Hoesen, and A. Damasio, "Severe pathological changes of the parabrachial nucleus in Alzheimer's disease," *NeuroReport* 9 (1998): 4151–54.

27 G. W. Van Hoesen and A. Damasio, "Neural correlates of cognitive impairment in Alzheimer's disease," in *Handbook of Physiology*, vol. 5, "Higher Functions of the Nervous System," ed. V. Mountcastle and F. Plum (Bethesda, Md.: American Physiological Society, 1987): 871–98; T. Grabowski and A. Damasio, "Definition, clinical features, and neuroanatomical basis of dementia," in *The Neuropathology of Dementia*, ed. M. M. Esiri and J. H. Morris (New York: Cambridge University Press, 1997): 1–20.

28 As we continue mapping the Alzheimer changes at different points in the disease, it will be possible to correlate neural sites and cognitive/behavioral defects more precisely, and this should be pursued actively since it is one of the few means we have to find answers to these problems. In all likelihood, the newly discovered Alzheimer pathology in the parabrachial nucleus will turn out to be contributory to part of the defect if not to all of it. It will most certainly be related to the autonomic dysfunction encountered in these patients and may even be a possible cause of their disproportionate incidence of respiratory and gastroenteric disease.

29 There is an intriguing suggestion that when glycogen stores located in glial cells are depleted by repeated neurotransmitter release, adenosine is released from glial cells and causes non-REM sleep to be induced. Non-REM sleep, in turn, allows glycogen to build up again in glia. See J. H. Benington and H. C. Heller, "Restoration of brain energy metabolism as the function of sleep," *Progress in Neurobiology* 45 (1995): 347–60.

30 For review, see B. Vogt, L. Vogt, E. Nimchinsky, and P. Hof, "Primate cingulate cortex chemoarchitecture and its disruption in Alzheimer's disease," in *Handbook of Chemical Neuroanatomy*, vol. 13, *The Primate Nervous System, Part I*, ed. F. E. Bloom, A. Bjorklund, and T. Hokfelt (New York: Elsevier Science B. V., 1997).

31 O. Devinsky, M. J. Morrell, and B. A. Vogt, "Contributions of anterior cingulate cortex to behavior," *Brain* 118 (1995): 279–306; P. Maquet, J.-M. Peters, J. Aerts, G. Delfiore, C. Degueldre, A. Luxen, and G. Franck, "Functional neuroanatomy of human rapid-eye-movement sleep and dreaming," *Nature* 383 (1996): 163–66; P. Maquet, C. Degueldre, G. Delfiore, J. Aerts, J.-M. Peters, A. Luxen, and G. Franck, "Functional neuroanatomy of human slow wave sleep," *The Journal of Neuroscience* 17 (1997): 2807–12; T. Paus, R. J. Zatorre, N. Hofle, Z. Caramanos, J. Gotman, M. Petrides, and A. C. Evans, "Time-related changes in neural systems underlying attention and arousal during the performance of an auditory vigilance task," *Journal of Cognitive Neuroscience* 9 (1997): 392–408; P. Rainville, B. Carrier, R. Hofbauer, M. Bushnell, and G. Duncan, "Dissociation of pain sensory and affective dimensions using hypnotic modulation," *Pain* (in press); P. Fiset, T. Paus, T. Daloze, G. Plourde, N. Hofle, N. Hajj-Ali, and A. Evans, "Effect of

propofol-induced anesthesia on regional cerebral blood-flow: A positron emission tomography (PET) study," *Society for Neuroscience* 22 (1996): 909; A. R. Braun, T. J. Balkin, N. J. Wesensten, F. Gwadry, R. E. Carson, M. Varga, P. Baldwin, G. Belenky, and P. Herscovitch, "Dissociated pattern of activity in visual cortices and their projections during human rapid eye movement sleep," *Science* 279 (1998): 91–95.

32 See A. Damasio and G. W. Van Hoesen, "Emotional disturbances," in *Neuropsychology of Human Emotion*, 1983; M. I. Posner and S. E. Petersen, "The attention system of the human brain," *Annual Review of Neuroscience* 13 (1990): 25–42.

33 Macdonald Critchley, *The Parietal Lobes* (London: E. Arnold, 1953).

34 Barry E. Stein and M. Alex Meredith, *The Merging of the Senses* (Cambridge, Mass.: MIT Press, 1993).

35 The rich interconnectivity of the superior colliculi led Bernard Strehler to suggest that they are, quite literally, the seat of consciousness. This is a rather extreme view and I am by no means endorsing it here. The hypothesis I present is entirely different, of course, but Strehler's review of collicular function was insightful. B. Strehler, "Where is the self? A neuroanatomical theory of consciousness," *Synapse* 7 (1994): 44–91.

36 E. G. Jones, "Viewpoint: The core and matrix of thalamic organization," *Neuroscience* 85 (1998): 331–45. We also know, from the work of E. G. Jones in primates, that neurons in several diffusely projecting thalamic nuclei (including intralaminar nuclei but not exclusively so), whose input comes from the brainstem tegmentum, have a specific chemical signature: calbindin. On the other hand, neurons in specific relay nuclei, whose input comes from lemniscal tracts and whose projection is topographically ordered, have a different signature marker: parvalabumin.

37 H. T. Chugani, "Metabolic imaging: A window on brain development and plasticity," *Neuroscientist* 5 (1999): 29–40.

38 A. Damasio, "Disorders of complex visual processing," in *Principles of Behavioral Neurology*, ed. M.-Marcel Mesulam, Contemporary Neurology Series (Philadelphia: F. A. Davis, 1985): 259–88.

39 Lawrence Weiskrantz, *Consciousness Lost and Found: A Neuropsychological Exploration* (New York: Oxford University Press, 1997).

40 A. Damasio, *Descartes' Error*; R. M. Brickner, "An interpretation of frontal lobe function based upon the study of a case of partial bilateral frontal lobectomy," *Research Publications of the Association for Research in Nervous and Mental Disease*, 13 (1934): 259–351; Richard M. Brickner, *The Intellectual Functions of the Frontal Lobes: A Study Based upon Observation of a Man after Partial Bilateral Frontal Lobectomy* (New York: Macmillan,

1936); Joaquin Fuster, *The Prefrontal Cortex: Anatomy, Physiology, and Neuropsychology of the Frontal Lobe*, 2nd ed. (New York: Raven Press, 1989). Note that I am not including the premotor cortices of areas 6 and 24 within prefrontal cortices, because they are functionally and architectonically distinct. Bilateral damage of premotor cortices is a rare natural event and has been difficult to investigate experimentally.

Chapter Nine: Feeling Feelings

1 These mechanisms were proposed in *Descartes' Error*, where they are discussed in detail.

2 See Vittorio Gallese and Alvin Goodman, "Mirror neurons and the simulation theory of mind-reading," *Trends in Cognitive Sciences* 2:12 (1998): 493–501.

3 The reader is directed to Jaak Panksepp's work on peptides connected to emotions for a review of this important aspect of background feelings. See J. Panksepp, E. Nelson, and M. Bekkedal, "Brain systems for the mediation of social separation-distress and social-reward: Evolutionary antecedents and neuropeptide intermediaries," *Annals of the New York Academy of Sciences* 807 (1997): 78–100; E. E. Nelson and J. Panksepp, "Brain substrates of infant-mother attachment: Contributions of opioids, oxytocin, and norepinephrine," *Neuroscience and Biobehavioral Reviews* 22 (1998): 437–52.

4 I thank readers of *Descartes' Error* for calling my attention to the work of Susanne Langer (*Philosophy in a New Key: A Study in the Symbolism of Reasons, Rite and Art* [Cambridge, Mass.: Harvard University Press, 1942]) and of Daniel Stern (*The Interpersonal World of the Infant: A View from Psychoanalysis and Developmental Psychology* [New York: Basic Books, 1985]).

5 G. W. Hohmann, "Some effects of spinal cord lesions on experienced emotional feelings," *Psychophysiology* 3 (1966): 143–56; P. Montoya and R. Schandry, "Emotional experience and heartbeat perception in patients with spinal-cord injury and control subjects," *Journal of Psychophysiology* 8 (1994): 289–96.

6 W. B. Cannon, "The James-Lange Theory of Emotions: A Critical Examination and an Alternative Theory," *American Journal of Psychology* 39 (1927): 106–24.

7 J.-D. Bauby, *Le scaphandre et le papillon* (cited earlier); J. Mozersky, *Locked In* (cited earlier).

8 J. L. McGaugh, "Involvement of hormonal and neuromodulatory systems in the regulation of memory storage," *Annual Review of Neuroscience* 12 (1989): 255–87; J. L. McGaugh, "Significance and remembrance: the role of neuromodulatory systems," *Psychological Science* 1 (1990): 15–25.

Chapter Ten: Using Consciousness

1 J. F. Kihlstrom, "The cognitive unconscious," 1987; A. S. Reber, *Implicit Learning and Tacit Knowledge* (cited earlier).

2 See Victoria Fromkin and Charles Rodman, *An Introduction to Language,* 6th ed. (New York: Harcourt Brace, 1997).

3 For the evolutionary background to the nonconscious knowledge of grammar see Steven Pinker's *The Language Instinct* (New York: Morrow, 1994); for the nonconscious nature of artificial grammars see Reber, *Implicit Learning* (cited above).

4 A. Bechara, D. Tranel, H. Damasio, R. Adolphs, C. Rockland, and A. Damasio, "A double dissociation of conditioning and declarative knowledge relative to the amygdala and hippocampus in humans," *Science* 269 (1995): 1115–18; S. Corkin, "Tactually guided maze learning in man: Effects of unilateral cortical excisions and bilateral hippocampal lesions," *Neuropsychologia* 3 (1965): 339–51; D. Tranel and A. Damasio, "Knowledge without awareness: An autonomic index of facial recognition by prosopagnosics," *Science* 228 (1985): 1453–54; A. Damasio, D. Tranel, and H. Damasio, "Face agnosia and the neural substrates of memory," *Annual Review of Neuroscience* 13 (1990): 89–109; L. Weiskrantz, *Consciousness Lost and Found* (cited earlier).

5 A. Bechara, H. Damasio, D. Tranel, and A. Damasio, "Deciding advantageously before knowing the advantageous strategy," *Science* 275 (1997): 1293–95; R. Adolphs, H. Damasio, D. Tranel, and A. Damasio, "Cortical systems for the recognition of emotion in facial expressions," *Journal of Neuroscience* 16 (1996): 7678–87; A. Bechara, A. Damasio, H. Damasio, and S. W. Anderson, "Insensitivity to future consequences following damage to human prefrontal cortex," *Cognition* 50 (1994): 7–15.

6 F. Jackson, "Epiphenomenal qualia," *Philosophical Quarterly* 32 (1982): 127–36.

7 Patricia Churchland has written a delightful discussion of the Mary thought experiment in "The Hornswoggle Problem," *Journal of Consciousness Studies* 3 (1996): 402–8.

Chapter Eleven: Under the Light

1 Nicolas Malebranche, *De la recherche de la verité* (Paris: A. Pralard, 1678–79): 914. "C'est par la lumière et par une idée claire que l'esprit voit les essences des choses, les nombres et l'etendue. C'est par une idée confuse ou par sentiment, qu'il juge de l'existence des créatures, et qu'il connaît la sienne propre" (author's translation). I thank Fernando Gil for calling my attention to Malebranche.

Appendix: Notes on Mind and Brain

1   A. Einstein, cited in J. Hadamard, *The Psychology of Invention in the Mathematical Field* (Princeton, NJ: Princeton University Press, 1945).

2   D. Hubel, *Eye, Brain and Vision* (New York: Scientific American Library, 1988). For background on selectional systems in biology see Jean-Pierre Changeux, *Neuronal Man: The Biology of Mind* (New York: Pantheon, 1985) and Gerald Edelman, *Neural Darwinism: The Theory of Neuronal Group* Selection (New York: Basic Books, 1987).

3   A. Damasio, "Time-locked multiregional retroactivation: A systems level proposal for the neural substrates of recall and recognition," *Cognition* 33 (1989): 25–62; A. Damasio, "The brain binds entities and events by multiregional activation from convergence zones," *Neural Computation* 1 (1989): 123–32; A. Damasio, 1994/1995 (cited earlier); G. Edelman, *Neural Darwinism* (cited above); R. Llinás and D. Paré, "Of dreaming and wakefulness," *Neuroscience* 44 (1991): 521–353.

4   W. Singer, C. Gray, A. Engel, P. Koenig, A. Artola, and S. Brocher, "Formation of cortical cell assemblies," *Symposia on Quantitative Biology* 55: 929–52.

5   F. Crick, *The Astonishing Hypothesis: The Scientific Search for the Soul* (New York: Scribner, 1994); F. Crick and C. Koch, "Constraints on cortical and thalamic projections: The no-strong-loops hypothesis," *Nature* 391 (1998): 245–50.

6   J.-P. Changeux, *Neuronal Man* (cited above); G. Edelman, *Neural Darwinism* (cited above).

ACKNOWLEDGMENTS

My first thanks go to Seamus Heaney for his unwitting contribution to the title of this book. The ending of his poem "Song" speaks of "when the bird sings very close to the music of what happens." *The Feeling of What Happens* was my spontaneous and perhaps inevitable adaptation of his verse to the specific topic of this book.

During the preparation of the manuscript I was fortunate to spend many hours discussing my ideas with knowledgeable and patient colleagues. I single out Hanna Damasio, whose thoughts and suggestions are a continuous inspiration; Josef Parvizi, whose expertise on the brain stem helped shape my views of this brain region and made the task of contending with its intricacies easier than it would have been without his enthusiasm; Ralph Adolphs, who keeps an open mind but never takes any explanation for granted; Charles

Rockland, who hardly ever accepts any explanation but is a most constructive and generous colleague; Patricia Churchland, whose insistence on crystal clarity is a welcome challenge; and my eternal critic Mrs. Lundy, who was far less severe than I expected. During this period I also had the advice of many colleagues who read the text and offered suggestions. They include Victoria Fromkin, Jack Fromkin, Paul Churchland, Fernando Gil, Jerome Kagan, Fred Plum, Pierre Rainville, Kathleen Rockland, Daniel Tranel, Stefan Heck, Antoine Bechara, Samuel Dunnam, Ursula Bellugi, and Edward Klima. I benefited immensely from their comments and I thank them for their wisdom and kindness.

I am equally grateful for the friendship with which, once the manuscript was completed, several colleagues thoroughly read and generously commented on the text. They are Gerald Edelman, Giulio Tononi, Jean-Pierre Changeux, Francis Crick, Thomas Metzinger, and David Hubel, who, as the reader of one's dreams, leaves no idea unexamined and no comma unturned. The responsibility for the errors and oddities that remain is mine, of course.

I am grateful to my colleagues in the Department of Neurology at the University of Iowa, especially to the members of the Division of Cognitive Neuroscience, for what they have taught me through the years and for the spirit with which they have helped create a unique environment for the investigation of brain and mind; and the National Institute of Neurological Diseases and the Mathers Foundation, whose grants have made that environment a reality. I am equally grateful to the neurological patients who have been studied in our Cognitive Neuroscience unit, for the opportunity they have given us to understand their problems.

My assistant, Neal Purdum, coordinated the preparation of the manuscript and both Betty Redeker, who has coped with my handwriting for sixteen years, and Donna Wenell typed the manuscript with professionalism and dedication. Denise Krutzfeldt and Jon Spradling helped me with bibliographical searches with their usual proficiency.

I acknowledge with gratitude the support and guidance of two friends, Jane Isay and Michael Carlisle, without whose advice and enthusiasm it would not have been possible to finish this project. Finally, I must note how enjoyable it has been to work with Ravi Mirchandani, a kindred spirit, in the preparation of the British edition of this book.

INDEX